Anti-Vaccination and the Media

Allison Cavanagh

Anti-Vaccination and the Media

Historical Perspectives

palgrave
macmillan

Allison Cavanagh
Clothworkers Building North 1.22
University of Leeds
Leeds, UK

ISBN 978-3-031-70558-8 ISBN 978-3-031-70559-5 (eBook)
https://doi.org/10.1007/978-3-031-70559-5

This Palgrave Macmillan imprint is published by the registered company Springer Nature Switzerland AG.
The registered company address is: Gewerbestrasse 11, 6330 Cham, Switzerland

If disposing of this product, please recycle the paper.

Acknowledgements

This work could not have been completed without the help and advice of many wonderful colleagues. I would here particularly like to thank Noah Leve and Callum Baldwin for their help, expertise and calm professionalism, and my family, especially my children Eliza and Sam, who make each day wonderful. Mum, every single day.

CONTENTS

LIST OF FIGURES

CHAPTER 1

Introduction

Abstract This chapter introduces the idea of biases of communication in respect of anti-vaccination discourse. The chapter provides an overview of the main issues around journalists' depictions of inoculation and opposition to it and lays the foundation for empirical work comparing two 'crisis' moments in the history of anti-vaccination in the UK, those around polio and subsequently those around pertussis/whooping cough. The chapter makes the case for considering anti-vaccination/vaccine hesitancy discourses as localised rather than universal and places emphasis on national concerns and nationalism as key sites for the production of meaning around inoculation.

Keywords Polio • Pertussis • History • Nationalism • Inoculation • Anti-vaccination

In her landmark study of opposition to the smallpox vaccine in the nineteenth century, Durbach argues that 'anti-vaccinationism' (was) 'at the very center of wider public debates over the extent of government intervention in the private lives of its citizens, the values of a liberal society, and the politics of class' (Durbach 2005: 6). This at least is as true of modern anti-vaccination discourse as it was of its earlier counterparts. It is always tempting in any historical analysis to draw a wider-ranging continuum between the past and the present, seeing what is true of one era as applicable to the present. It is tempting to argue that what was true of smallpox can be applied readily to the case of Covid, and other modern vaccination conundrums. The groups affected may have altered, and the profiles of the afflicted changed, but the imperatives seem the same. At a time of global need, and few would argue that the recent pandemic is not an example of this, reaching into the past for solutions to the present is only natural. The 'lessons of history', we reason, should provide us with guidance (or what, we feel, is history for?). Even if the past cannot provide answers, that basic sense of security that comes from knowing similar instances which have been faced before and overcome is not to be understated. The problem with this comforting tendency though is that it is not helpful. The past will not offer up a recipe for handling the problems of the present. Histories do however offer a way to interrogate and unsettle assumptions. If we cannot use the past to divine answers, can we at least hope to frame more pertinent questions? I think so, at least insofar as histories prevent matters of ideology coalescing into fact. One example of this is the belief common in popular discourse and entirely unfounded in historical fact that anti-vaccination is an 'American' phenomenon. As many historians have pointed out, and this study also reiterates, the UK was and remains at the centre of the production of discourses of vaccine hesitancy in their substance. It is true that the work of circulating 'anti-vax' misinformation, especially on social media, is largely performed elsewhere (see Ong and Cabanes 2019), but the themes and rhetoric are all 'home grown'.

Of course, the nature of vaccine refusal could hardly be posed more urgently. In a world which has passed an 'event horizon' of global climate collapse, the proliferation of new viruses and the return of old foes, only recently sent packing by advances in public health and immunisation, is certain. The fallout of global heating, mass dislocation and human and animal migration cannot but prove fertile ground for a new virological ecosystem. Inequalities in access to healthcare globally continue to confound the best efforts to universalise access to immunisations with the additional burden of startling declines in childhood coverage since 2019 (see Dadari et al. n.d.). Even mature and well-established vaccination programmes can fall victim to their own success as first-hand awareness of the effects of viruses recedes from public memory. The shock and dismay witnessed in the West during the Covid-19 pandemic is itself a testament to how far our expectations of health have risen.

As the historian Simon Schama has recently reflected, much of the reaction to Covid in the west smacks of a dawning awareness of peril.

Perhaps it is chagrin at the realisation that the best-laid plans of mice and monsters are so many vanity projects compared to the entropy of the habitable planet, or the eruption of pandemics, that makes for reluctance to describe those existential crises in anything but the stale vocabulary of political and military history. Diseases are invaders; measures to deal with them a plot; bacteriologists and epidemiologists an alien elite, the microbe and the scientist in cahoots against homespun wisdom. The heath of the world contracts into the health of nations even when the latter cannot be sustained without ensuring the former. Much madness has arisen from this ongoing drama of false consciousness; and many are the perils arising from its obstinate perpetuation. (Schama 2023: 8)

This chagrin is also tied up with a rough-and-ready form of anti-scientism, in many ways itself a reaction to this ongoing re-evaluation of humanity's place in the natural world. As the promises of mastery of nature offered by techno-utopian fantasies of the twentieth century retreat in the face of the realities of the twenty-first, disillusionment manifests as petulant wrath. In both the US and the UK, the populist right has embraced anti-science, rebranding ignorance as scepticism and nihilism as a fatuous badge of honour. On the left, likewise, a weak relativism which raises the anecdotal and the idiosyncratic to the status of knowledge and repudiates scientific intercession has been damaging. Taken together the politicisation of science has come to be such that advances thought to be at least 'birds in the hand' are in danger of being lost altogether.

THE BIASES OF VACCINATION COMMUNICATION: CONTROVERSIAL, CONTEMPORARY AND COMPELLING

Vaccine hesitancy, as it is now described, has been identified by the WHO as one of the key threats to global public health, a definition which encompasses 'delay in acceptance or refusal of vaccination despite availability of vaccination services. Vaccine hesitancy is complex and context specific, varying across time, place and vaccines. It is influenced by factors such as complacency, convenience and confidence' (MacDonald 2015: 4163).

Communication around anti-vaccination was identified by the WHO SAGE group as a key component of delay or refusal, but not a determining one. The group reviewed grey literature on communication examining vaccine uptake, concluding that dialogue-based strategies, especially those informed by the use of community leaders to establish trust and channel communications effectively, along with the use of recall/reminder prompting communications, and non-financial incentives to attend immunisation clinics were key to addressing this communications deficit. What is notable here is the exclusive focus face-to-face communications, the assumption being that physical co-presence is the requirement of persuasive communications. The report refers to social media, as organising environments for communication, but relegate mass media to that of a 'back up communication system' to target populations with low awareness of health services. Yet they acknowledge that

> mass vaccine promotion campaigns may enhance positive attitudes towards vaccination and, ultimately, increase coverage rates. However, interventions using mass media are difficult to evaluate and are not well-suited to experimental design; other types of evaluation are subject to various forms of bias due to the many potential confounding factors which limits the quality of the evidence available. When communication interventions are part of multi-component strategies, it becomes almost impossible to evaluate their direct impact on vaccine uptake. (Dubé et al. 2015)

Communication in the paradigm of health communication is understood as the effective transfer of 'correct' information through the filters of social obstacles. As Leach and Fairhead have argued, this is fundamentally problematic. 'It is not enough', they explain 'simply to draw a contrast between science/policy and public framings, or between globalized and personalized ones, as if they were part of distinct, separate lifeworlds. Rather, crucial questions concern how these contrasts arise, become manifested and consolidated' (2007: 4).

To begin to unpick some of these, we should start by observing that the literature on anti-vaccine communications is subject to systematic biases towards a set of themes which inform both the work of stakeholders, policymakers, practitioners on the one hand and the vaccine hesitant on the other. These common concerns and assumptions structure a reading of the issues which is shared across the community, albeit with a different value inflection. I will consider each of these in turn.

THE NEW!

Firstly, as noted in Chap. 2, by far the majority of articles and texts published about vaccine hesitancy start from the position that social media is the most significant point of reference and subject matter. This is in part a function of the ease with which social media materials can be accessed and their visibility for analysts and researchers. New methods of representing patterns in 'big' data further make for a greater accessibility in findings and a greater ease in translating the, sometimes seemingly esoteric, concerns of academics into the language and genres of communication of policy stakeholders. On the other hand, the reverse is also true. Methods which are used to render social media materials into a format which is 'ready to hand' for the same stakeholders do not translate easily into other forms of public representation and mediation, often requiring accompanying explanations of method which are a struggle to convey in the attenuated language required in public communications. That, however, is true of any social issue and not unique to health communications or anti-vaccination materials.

Social media, on the other hand, is also one of the few arenas that can act as a conduit specifically for anti-vaccination materials. We can draw a comparison between the way those who are both opposed to vaccination and keen to discuss and/or proselytise are able to use social media, and the forms of written materials which circulated in early anti-vaccination movements. A good example here would be Tebb's *Vaccination Inquirer*, the journal of the London Society Abolition of Compulsory Vaccination in the 1880s. Such types of publication, including personal letters, letters to the editor, pamphlets and periodicals, were designed to speak only to a community already in agreement with the movement, and in themselves form a means of reconfirming community beliefs and reinforcing cohesion. This of course is also true of social media in the modern age. Discussion of anti-vaccination, or expressions of vaccine hesitancy in generalised discussions online, acts to assemble an audience of those who already hold these views or concerns to a greater or a lesser extent. This provides us with some grounds for tracing the genealogies of anti-vaccination media in this way. However, anti-vaccination publications in the nineteenth century were not alternatives to other media in the way we would consider this today, and indeed those who were part of the addressed would also have had an extensive engagement with other special and general interest periodicals and publications. In comparison, social media communities of anti-vaccination advocates in our own era are often characterised by their inherent distrust of 'mainstream' media and a greater tendency to seek alternative forms of information. In looking at social media, analysts feel that they have found the 'smoking gun' of persuasive rhetoric around anti-vaccination.

This assumption, however, overlooks the very significant role of mainstream media in the development of these vaccine-hesitant narratives. The 'spikes' in discussion of immunisation concerns in the mainstream media, even in the publications considered in this study, are often correlated with the broadcast of impactful news items or documentaries which resulted in a public concern. Since these outbreaks of concern brought in their wake a short-lived surge in news, editorials, letters and responses, they are often credited with initiating change for better or worse in parental attitudes to vaccines. Moreover, the importance of newspaper and mass media campaigns in the history of health communication cannot be overstated, and anti-vaccination is no exception. There is a long and lively tradition of advocacy journalism in the history of health and health scandals, especially where there is an additional moral or consumer focus to the problem. We can also trace the history of the 'campaigning newspaper' back to the second half of the nineteenth century and the coming of an aspiration to 'government by journalism' (Stead 1888; see Chap. 5 for further discussion).

The historian Simon Schama has recently pointed to how ironic it is that the proliferation of modes of communication during the Covid-19 pandemic served to underscore the precarity of mass communication more generally. Contrasting the recession of human presence from urban spaces with the reinvigoration of nature, he points to the 'empty desolation of cities, the grim, still, silence of locked-down streets and squares'. 'While we hunkered and cowered, and ordered home delivery, flora rioted; fauna, trespassed. Parliaments of legislators were reduced to socially distanced barking from the hollow shell of their chambers … the more we retreated into digital numbed companionship, the more brazenly the company of animals advanced towards us' (2023: 13). In similar terms, we can note that social media proved to be a fertile ground for anti-vaccination and vaccine-hesitant narratives during Covid to the extent that this isolation continued. The greater the remove from access to a democratic public sphere, the more likely the proliferation of opposition (see below and Chap. 5), to which we now turn.

THE OPPOSED! (AND THE REST)

It is a commonplace amongst any who have ever worked in a public-facing role that feedback and complaint are most usually practical synonyms. The phrase 'no news is good news' could have been coined for use in handling public feedback, especially as this appears in the context of consumption. As McKeever et al. (2016) discovered from their survey of mothers in the US, those who opposed immunisation for their children were also more likely to engage in communications activities than those who did not. McKeever et al. applied Noelle-Neumann's model of the spiral of silence in part to explain this. In essence, the spiral of silence approach accounts for the way that perceptions of the views of a community will influence participation within it. Noelle-Neumann argued that the further away from the perceived 'average' opinion of a group an individual interlocutor perceived themselves to be, the lesser the likelihood that they would express an opinion on a matter of common concern. This leads in turn to a 'spiral of silence' insofar as groups' perception of where this 'average' opinion is will become increasingly rarefied as dissenting and qualifying voices self-censor. The process is self-perpetuating as each round invokes a further iteration of self-exclusion as the identified 'average' of the group continues to move towards more extreme positions. This mechanism is one of the most useful in accounting for the often-rapid movement of groups towards extreme positions, as each 'moderate' voice begins to feel the chill of exclusion (see Noelle-Neumann 1974). However, McKeever et al. argue that there is something of a self-limit on the process in the case of social issues. Although the spiral of silence approach is of noted utility in the case of political communication where there is a continual renewal and change in issues and the signifying elements within them, in social issues the gradual drift to extreme views can result in an energetic backlash and radicalisation of 'outsiders' where they perceive their values to be under threat (2016: 482). It is also likely in such cases that the perceived 'consensus' is actually a 'straw man'. In the case of the mothers studied by McKeever et al., those who were pro-vaccination or ambivalent tended to communicate less than those who were vaccine hesitant, and thus the availability of a 'reality check' exposure to alternate views is compromised.

As with individuals, so with mass media. The story of successful and safe vaccines which have been working effectively for years is simply not a story at all. Indeed, if our vaccines, mostly administered in childhood, are effective, they will not occupy our attention to any degree. Illnesses averted are, unlike other risks and hazards, not experienced as a near miss so much as not experienced at all. The growth of health and wellness industries continues to reaffirm deeply held societal beliefs that health is a behaviour that can be cultivated (see Ehrenreich 2010). As a result, we are more likely to credit not having developed a virus to our own wise lifestyles than to our parents' health decisions on our behalf. Where risk is invoked in respect of vaccines, the scales are stacked on one side only, the risk of taking action. People will tend towards what health practitioners may see as 'complacency', or alternately as risk aversion, where the threat from the disease has receded, ironically itself often a function of herd immunity.

If 'vaccine still working' is not news, then how else can the scales be balanced? In theory through a focus on news stories of those who are afflicted by a given illness and also un-vaccinated. However, even in the jaded world of modern news production, what news outlet would likely see fit to publish a story which offered up an individual victim of a virus with life-altering or worse prognoses alongside a censorious footnote that the victim was unvaccinated? To present the story of the success of immunisation, the press can only have recourse to pale statistics, which jar with the 'emotion work' of the professional journalist (see Wahl-Jorgensen 2019), or rely on newsworthy briefs from scientists and policymakers which might transform the everyday into an event and render the benefits available visible. This, however, returns us to the points raised above concerning the populist distrust of science.

In order, then, for vaccination stories, whether lay or expert, to achieve a balance of exposure there must be some additional anchor or framing which renders it visible—in short, the compelling.

THE COMPELLING!

As the WHO have noted (see above), vaccine hesitancy is articulated through and comes to public attention as a result of a cycle of communication which moves through complacency about the risks of the illness to more active engagement with immunisation campaigns but does not necessarily do so in accordance with the prevalence of illness within communities. On the contrary, outbreaks of serious illness can, ironically, lead to a lower uptake of vaccines for other infections, as people's trust in healthcare settings themselves becomes compromised. Publics may become more anxious about immunisation or more militant in the face of disasters, more questioning in the case of outbreaks. A good example here is that of the 2014 Disneyland measles outbreak. At this time, the US was experiencing a high level of measles cases in the general population, 667 cases, the highest reported over the previous 20 years (Doll and Correira 2021). The Disneyland event became a news item with one case of an unvaccinated Californian child who was hospitalised with measles after a visit to Disneyland. Additional cases were identified within days and the infection 'toll' for this event was settled at 125 by the CDC (ibid.: 4210). The source of the outbreak was later traced to Canada, and in both the Canadian and US cases, the profile of the infected included a refusal of immunisation based on religious or other belief. The high level of public knowledge of the outbreak was credited with driving an increase in positive social media posts and shared content concerning vaccination in the immediate wake of the event. Doll and Correira also suggest that the event acted a trigger which mobilised a community less invested in immunisation. Although the evidence gathered in this study is tentative, the link between the high-profile media discussions of vaccination and more generally vaccination-positive sentiments was validated. That this outbreak achieved such a high media profile however is the outcome of the combination of rising measles cases in the immediately preceding period and the media-friendly pathos inherent in the sad case of a child contracting a serious illness in the 'happiest place on Earth'. It is credited with smoothing the passage of California's (2021) bill to remove vaccine exemptions for public school children on the grounds of personal belief.

Outbreaks of disease are one of the few newsworthy occasions where vaccine-positive beliefs are shown alongside vaccine-hesitant or anti-vaccination views in the media. They come, though, with an already-virulent set of moral attributions. There can be few arenas of collective life more amenable to moralisation and more contested than outbreaks of disease, as I will argue in Chap. 2. Thus, for the current purposes we can see that there is a systematic bias in communication about vaccines, across both lay and scientific communities, towards immediacy, negative news and innovation.

OUTLINE OF THE CHAPTERS

In this chapter, I will be looking at the way in which models of representations of health and illness can be used to examine the representation of inoculation. I'm starting by looking at why representation has been of particular interest in the context of health and how this stream of research was established. This beats a path through works focused on framing, metaphor and representation, critical discourse analysis and social representations. I here argue that theorists have used the representation of illness as a way to access purported deeper structures of meaning underpinning social representations. These anthropological approaches to representation have served to reveal the extent to which politicised narratives of nationality, identity and belonging have come to be attached to sickness. Along the way, however, such themes necessarily foreground questions of blame, guilt and the nature of citizenship in modern societies. The extent to which people owe obligations in respect of health to the state, or at least an obligation of conformity to those mechanisms deemed necessary for its pursuit are examined here.

In Chap. 2, I will be going on to look at a second set of approaches, those informed by 'big data' perspectives to historical studies. There is good reason why these more quantitative approaches are coming to play an increasing role in historical studies within the academy. I trace here the forms in which both big data and its cousin relational data analyses enter into the range of historical work and trace some of the effects of this on the field as a whole. I argue here that questions posed by the availability and analysis of big data are of a unique character, one which serves to foreground partial aspects of analysis and which can only be demonstrated, confirmed or rebutted on the basis of alternative big data, or reinterpretations of the same data set. On the one hand, this might not appear immediately controversial, but, as I argue in this chapter, the nature of the collated data in the case of 'big data' makes comparability of alternative data sets problematic. On the one hand, if a data set covers the entirety of records of a given historical phenomenon, then an alternative data set is not available; on the other, as I argue below, any given data set is comprised of both the data/texts and the assumptions which have structured the assemblage, for example about what materials can be rightly treated as a unity/equivalency. Chapter 3 considers the affordances and consequences of this approach for a study such as this.

Chapters 4 and 5 examine two case studies from the wider data set. Chapter 4 looks at polio in the period 1955–1960. The archive contained all references to polio in the daily newspapers *The Daily Mail* and *The Daily Telegraph* for this period. These were used to generate an archive accessible to input. The reports were 'cleaned' using the Terrier stop list and corrected for errors based on degraded typography. One of the challenges of working with digitised newsprint archives is monitoring the effectiveness of the optical character recognition (OCR) bundled into the platform. In the case of these, the decay was minimal and the OCR effective. These materials were then converted into networks using the Visone graphical user interface (see Brandes and Wagner 2004), to determine the relationship between clusters of signifying elements (see below for further discussion). These were then compared, and they form the basis of the quantitative and relational analysis of the discussions. Chapter 5 is devoted to the case study of the whooping cough vaccine in the UK during the period 1974–1979, using the same newspapers and methods. Using this combination of qualitative and quantitative data, specifically semantic network analysis, I examine how the papers reported on vaccination programmes and further the ways that vaccine hesitancy was depicted and incorporated into the reportage. The aim here is to uncover discourses underpinning opposition to or hesitancy over immunisation and how this altered between these two 'crisis points'.

In identifying these, I am using Tarrow's notion of cycles of contention/activism as a jumping-off point. Tarrow identifies cycles of contention as 'increasing and then decreasing waves of interrelated collective actions and reactions to them whose aggregate frequency, intensity, and forms increase and then decline in rough chronological proximity' (Tarrow 1993: 287). Tarrow goes on to apply Tilly's concept of the repertoire of social action to the case of cycles of contention with the aim of establishing why it is that the techniques of protest and expression of concern within a society exhibit remarkable stability over time. Repertoires here refer to the sum of the skills, rhetoric, communicative forms and cultural understandings within a society to which protesters can have recourse and which determine the viability and the legitimacy of their protests. At certain historical junctures, these will be subject to change, what Tarrow terms 'moments of madness', in which the 'right' ways of 'doing' protest or the right issues on which to protest seem to change. Tarrow argued that few of these survive their initial outing in the public sphere:

> Moments of madness-seldom widely shared, usually rapidly suppressed, and soon condemned even by their participants-appear as sharp peaks on the long curve of history. New forms of contention flare up briefly within them and disappear, and their rate of absorption into the ongoing repertoire is slow and partial. But the cycles they trigger last longer and have broader influence than the moments …themselves. (1993: 302–3)

Such is the case with anti-vaccination narratives, considered as part of an engagement with the state on the rights and responsibilities of citizens with respect to health, as I will argue below. Using these insights we can identify several key crisis points in the history of immunisation in the UK. These begin, as outlined by Durbach (2005), with the case of small-pox vaccination in the nineteenth century, and go on to encompass tuberculosis (TB) (after 1953), the HPV vaccine introduction (2008) and, of course, the long shadow of the measles, mumps and rubella (MMR) debate, which developed from Andrew Wakefield's ill-starred publication in the Lancet in 1998. This work will compare two key moments in this longer narrative, moments sufficiently removed to allow for meaningful comparison but also sufficiently alike in terms of the environments of their emergence to be able to trace the legacies between them.

In Chap. 6, I will outline the significance of these findings and resituate them in the context of Covid-19. The Covid pandemic represents a point at which the nature of the pro- and anti-vaccination discourse becomes enmeshed in a broader and more diverse discourse of contention, claims and counterclaims. I will trace here the ways in which academics have sought to capture this transformation and how we can begin to incorporate this, now still raw for many, crisis event into a broader understanding of the nature of anti-vaccination movements and citizen participation in health communication.

References

Brandes, U. and Wagner, D. (2004) 'Analysis and Visualization of Social Networks', in M. Jünger and P. Mutzel (eds) Graph Drawing Software. Berlin, Heidelberg: Springer Berlin Heidelberg, pp. 321–340. Available at: https://doi.org/10.1007/978-3-642-18638-7_15.

Dadari, I., Belt, R., Iyengar, A., Ray, A., Hossain, I., Ali, D., Danielsson, N. and Sodha, S.V. (n.d.) 'Achieving the IA2030 Coverage and Equity Goals through a Renewed Focus on Urban Immunization', Vaccines, 11(4), p. 809.

Doll, M.K. and Correira, J.W. (2021) 'Revisiting the 2014–15 Disneyland Measles Outbreak and its Influence on Pediatric Vaccinations', Human Vaccines & Immunotherapeutic, 17(11), pp. 4210–4215.

Dubé, E., Gagnon, D. and MacDonald, N.E. (2015) 'Strategies intended to address vaccine hesitancy: Review of published reviews', WHO Recommendations Regarding Vaccine Hesitancy, 33(34), pp. 4191–4203. Available at: https://doi.org/10.1016/j.vaccine.2015.04.041.

Durbach, N. (2005) Bodily Matters: The Anti-Vaccination Movement in England, 1853–1907. Durham and London: Duke University Press.

Ehrenreich, B. (2010) Smile or Die: How Positive Thinking Fooled America and the World. London: Granta.

Leach, J. and Fairhead, M. (2007) Vaccine Anxieties: Global Science, Child Health and Society. London and New York: Routledge.

MacDonald, N. (2015) 'Vaccine hesitancy: Definition, scope and determinants.' Vaccine, 33, 34, pp. 4161–4.

McKeever, B.W. et al. (2016) 'Silent Majority: Childhood Vaccinations and Antecedents to Communicative Action', Mass Communication and Society, 19(4), pp. 476–498. Available at: https://doi.org/10.1080/1520543 6.2016.1148172.

Noelle-Neumann, E. (1974) 'The spiral of silence a theory of public opinion', Journal of Communication, 24(2), pp. 43–51.

Ong, J.C. and Cabanes, J.V.A. (2019) 'When Disinformation Studies Meets Production Studies: Social Identities and Moral Justifications in the Political Trolling Industry', International journal of communication (Online), p. 5771.

Schama, S. (2023) Foreign Bodies: Pandemics, Vaccines and the Health of Nations. New York: Simon & Schuster (epub.).

Stead, W. (1888) 'Government by journalism', Contemporary Review, (49), pp. 653–657.

Tarrow, S. (1993) 'Cycles of Collective Action: Between Moments of Madness and the Repertoire of Contention', Social Science History, 17(2), pp. 281–307. Available at: https://doi.org/10.2307/1171283.

Wahl-Jorgensen, K. (2019) Emotions, Media and Politics. Cambridge: Polity.

CHAPTER 2

Metaphor and Representation in the Mediation of Illness

Abstract This chapter introduces the main approaches which have commonly been used to study representations of illness and makes the case for their equal applicability to anti-vaccination discourse. The chapter provides an overview of studies of metaphor in illness, in particular tracing the influence of the work of Susan Sontag on the one hand and that of Sergei Moscovici and his exponents on the other. The chapter contrasts these thematic and metaphorical approaches with the forms of analysis which derive from critical discourse analysis and offers a pathway towards integrating their concerns through consideration of modalities. The chapter analyses the 'stories we tell' about illness and vaccination and traces their similarities and points of divergence with conspiracy theories.

Keywords Sontag • Representation • Moscovici • Illness • Theoretical approaches • Metaphor

INTRODUCTION

This chapter looks at the role of media representation in understanding how anti-vaccination narratives have developed over time. Little attention has been paid to this specifically in the literature on health and representation, understandably enough. The majority of work in the area has focused

A. Cavanagh, *Anti-Vaccination and the Media*,
https://doi.org/10.1007/978-3-031-70559-5_2

on examining the legibility and ease of interpretation of representations of immunisation campaign materials. As noted elsewhere in this work, this starting point is one which enshrines a naïve and overly simplistic view of the nature of communication in which clear divisions are posited between message and misunderstanding, signal and noise. The aim of these studies is to assess the barriers to either comprehension or alternatively to belief. Such studies take as their cue from the premise that representations can and do form obstacles to healthcare providers in indicating a rational line of action, and / or to patients in accepting the same and acting on it. This features a basic belief that if only the correct representational 'levers' and techniques can be deployed, naturally enough a vaccine rollout will achieve a high if not universal uptake. Such an approach though fails to grasp that the challenges of representation are seldom so simple or so unequivocal. The 'decisions' to take, hesitate or refuse (as the Report of the SAGE Working Group on Vaccine Hesitancy makes clear, refusal and hesitation are not the same thing, though the outcome may ultimately be (Nuwarda et al. 2022)) a given immunisation are not made solely on grounds of belief or attitude but are tied, as will be argued below, to community identification. Thus, a representational 'fix', howsoever well-conceived, may entirely miss the point as far as those who are the subject of these are concerned. Neither has much scholarly work attempted to grapple with the question of the nature of representation from within the anti-vaccination narrative. A notable exception here is the work of Fasce et al. (2023), whose impressive taxonomy of anti-vaccination arguments is discussed in detail in this Chap. 2. By and large, however, such systematic engagements are uncommon, and it is very unusual for anti-vaccination narratives to be treated in the same terms and subjected to the same analytical procedures as pro-vaccine materials. Most theoretically grounded approaches to understanding such communications start with literatures around conspiracy theories and/or consider how anti-vaccination concerns feed into a broader political or cultural setting. Most recent work has revolved around the attempt to 'track and trace' the circulation of such narratives and communications (work which again is discussed in Chap. 3) through social media systems using the availability of techniques of 'big data' and relational data analysis. Such an approach has tended to intensify a focus on the analysis of connections rather than the analysis of content which leads ultimately to a systematic imbalance in the field as a whole. Whereas studies from a healthcare perspective concern the use of representational techniques or rhetoric with the aim of eliminating barriers to 'rational' choices, studies of anti-vaccination narratives are more focused on the embedded nature of the communications, the political ends to which they are put. Neither approach has much to say about the content of the communications themselves.

To start to unpick this, then, we need a path to follow. There is a robust and extensive history of the analysis of metaphor in understanding the contexts and social constructions of illness, and this literature is distinct in its understanding, following Foucault, that rationality cannot be ascribed from outside of discourse but is instead a product of it. Thus, the ways in which people understand medical interventions and illnesses themselves are not to be seen as something independent of the 'real' nature of risk and threat, and neither is 'the voice of reason' a property or attribute of one party within the discourse. In this work, I will be arguing that these analyses of representation offer a greater analytical potential in uncovering the social meanings of vaccination and the manner in which anti-vaccination activists and the vaccine hesitant understand their experiences and rally a public of common interest. This is central to understanding not only why (rather than how) such discourses achieve widespread circulation and currency within modern social media dense societies but further how anti-vaccination narratives themselves appear and are developed.

To begin with, then, this chapter acts as a critical review of approaches and traces their development. I will begin by tracing and comparing the different approaches and considering how they foreground alternative understandings of the role of representations in public understandings of disease. I open with the work of Susan Sontag, moving on to look at social representations theory through the work of Sergei Moscovici, and conclude by looking at the work of Fairclough and aligned theorists in the field of critical discourse analysis especially the importance of modality in reinforcing metaphors and signification in anti-vaccination narratives.

SONTAG AND HER LEGACY

When Susan Sontag published her, now seminal, study of the ways in which the form of representation, metaphors and symbolism which had become attached to cancer served to stigmatise and marginalise those suffering from the disease, academic reaction was muted. Her work was seen as occupying ground somewhere between literary studies and psychology and was interpreted as an attempt to intellectualise her own cancer's 'emotional baggage'. Reviewers picked up on the status of Sontag as an insider within the cancer community, someone who was as a result of having cancer herself highly motivated to engage in its demythologisation. Demythologisation was seen to be a way to use 'literary means to achieve antimetaphoric and antihermeneutic ends' (Davis 2002: 828), to 'calm the imagination, not to incite it' (Sontag 1989: 14). For Sontag, the experience of seeing her 'fellow patients...in the grip of fantasies about their illness by which I was quite unseduced' (ibid.: 12), 'spectacularly ...encumbered by the trappings of metaphor' (ibid.: 5) was the spur to this work. The author was appalled by the idea that those who feared that they had cancer would avoid diagnosis, and/or hide or deny cancer diagnoses, for fear of the public stigma which attached to the disease. For critics, on the other hand, Sontag's status as a sufferer appeared to undermine her work, its apparent subjectivity consigning it to an intellectual hinterland between academia and journalism at a time when strenuous efforts were being made to place cultural studies of representation on a firm theoretical footing. For this reason, uptake of her ideas in the late 1970s was initially slow.

The ideas she developed in this work found greater purchase, and greater salience, when they were reapplied to the AIDS epidemic of the 1980s. By this point the stigma of this disease was balanced by a public awareness of the health dangers posed by said 'emotional baggage'. That cancer sufferers could suffer alone (and die unnecessarily) through fear of embarrassment and public shame was concerning; the fear that a like stigma might drive infection in a disease with no cure focused minds. In *AIDS and its Metaphors*, Sontag outlines what she describes as the 'plague metaphor' and shows how AIDS came to be constructed through its lens. To understand an illness as a plague is, as many have observed, to moralise it, and to *other* it (see Herzlich and Pierret 1987). Plagues, historically associated with a judgement on behaviour, become inflected by racialised and xenophobic fantasies (Sontag 1989: 48–50). The military metaphors of illness as a battleground, and concomitantly the patient as a combatant, which she had identified as the master frames in discourse about cancer, gives way with AIDS, Sontag argued, to metaphors of pollution and moral corruption. 'Cancerphobia', she claimed, 'taught us a fear of a polluting environment; now we have the fear of polluting people' (ibid.: 73). It is this fear which prompts a flight into introversion and a rejection and pathologising of altruism.

Her work has been widely used as a basis for a thematic and metaphorical analysis of illness and contagion, framing 'our understanding of the relationship between disease metaphors and illness experience in modern

Western society' (Clow 2001: 293; see also Kleinman 2020) and finding fertile ground in analyses of sufferers, policymakers and wider societal narratives. Most recently, her work has received renewed interest as academics sought to respond to the panicked proliferation of discourse around the Covid-19 pandemic. Media scholars, sociologists and epidemiologists alike found utility in her theorisation of representations as a resource to assess metaphors of the pandemic and to consider the ways these informed, for better or worse, public and policymakers' responses to the coronavirus (see Herzlich and Pierret 1987; Hanne et al. 2007; van der Geest and Whyte 1989; Potts and Semino 2019; Brandt and Botelho 2020; Craig 2020).

Sontag compared the way that different illnesses, cancer and tuberculosis, acted as 'master illnesses', signifying and substituting for particularly entrenched social ills.

> Illnesses have always been used as metaphors to enliven charges that a society was corrupt or unjust. Traditional disease metaphors are principally a way of being vehement; they are, compared with the modern metaphors, relatively contentless…Particular diseases figure as examples of diseases in general; no disease has its own distinctive logic. Disease imagery is used to express concern for social order…Master illnesses like TB and cancer are more specifically polemical. They are used to propose new, critical standards of individual health, and to express a sense of dissatisfaction with society as such. (Sontag 1977: 72–3)

Master illnesses are seen to resonate with prevailing social conditions. In the industrial era, the supposed 'romantic' death from tuberculosis expressed and symbolised prevalent fears of social contagion and proximity unavoidable in rapidly urbanising Western societies. In the mid-twentieth century, with the valorisation of science and reason and its implied links with progress and a new political order emerging to soothe the horror of the old, the symbolism of cancer was linked to a loss of control (Small et al. 2006; Stacey 1997). However, the condition still serves as a master illness, howsoever inflected, providing a symbolic template for the representation of disease more generally.

Following in Sontag's footsteps, academics have sought to use mass media, and most particularly journalistic accounts to develop an understanding of the narratives and symbolic universes of illnesses. Wallis and Nerlich (n.d.) applied Sontag's approach to UK newspapers and the framing of the 2003 SARS outbreak, whilst Trčková (2015) considered metaphors of Ebola in the US press. Hanne and Hawken's comparative study of the representation of five illnesses in the *New York Times* (2007) concluded that whilst the metaphors themselves were unevenly distributed, the 'claim that human beings cannot talk about health and sickness without metaphor was amply borne out... metaphors used in the mass media play a significant role in shaping popular conceptions of sickness' (2007: 98). Other studies have sought to compare the prevalence of metaphors between national news outlets (e.g. see Chopra and Doody (2007) on the prevalence of metaphors of cancer and schizophrenia in the UK and US press). Williams Camus (2009) narrowed the focus of study to the popularisation of scientific articles in considering the role of science journalists as cultural intermediaries.

Applications of Sontag's work have not, however, been restricted to analyses of mass media content. Potts and Semino (2019) recently extended Sontag's work to a quantitative metaphorical analysis of three large English language corpora, concluding that, just as social evils are used to depict and demonise cancer, so the metaphor of cancer 'might be used to legitimize extreme measures as solutions to perceived threats' (2019: 94). Historical and comparative accounts, for example Patterson's work on the history of cancer in the US (Patterson 1987) and Rothman's 1995 study on tuberculosis, have also used Sontag's work as a jumping-off point.

Likewise, for some, recognition of the role played by metaphors in patients' understanding of their illness is essential to provision of care.

> Medicine is largely about storytelling and interpretation, and narrative, metaphor, and symbol are fundamental tools of the trade...Patients understand their illnesses in a narrative way whether their physicians realize it or not. If this is so, and if physicians ignore or devalue narrative, then health care is bound to suffer. (Coulehan 2003)

Of course, the criticisms of Sontag's work still stand to some extent. As many commentators have observed, there is a lack of empirical warrant for many of Sontag's claims. Her study is gestural and impressionistic with no real systematic underpinning. For Clow, moreover, there was little evidence of a transfer between the symbolic presentation of cancer in popular culture and the life world of cancer suffers. Examining a range of sources, including letters and obituaries, Clow concluded that 'neither shame nor silence were universal features of the cancer experience' (Clow 2001: 304) and, tellingly, that far from the conspiracy of silence around the disease identified by Sontag in the 1940s, this period saw a large proliferation of representation and discourse around cancer (Clow 2001: 300). Medical practitioners and cultural analysts alike have also contested the prevalence as well as the benefits and failings of the illness as war metaphor (see Reisfield and Wilson (2004); Clow (2001)).

Apart from questions of accuracy, there are serious omissions in Sontag's work. One key area of concern left relatively unexplored by Sontag is why illness itself acts as a locus for signification in this way. The cultural politics of illness are complex in their right, expressive of the boundaries of group identification, especially in the case of plague and infection. Why illness, regardless of its nature, should stand proxy for social conscience, a critical awareness of a society's own problems, is not developed in her account.

One answer to this may be found in seeing metaphors of illness as products of a wider process of social storytelling. For Stacey, 'stories about illness are an intensification of the way in which we generally understand our lives through narrative' (1997: 8). We understand our lives through stories, of self, of origins, of projections into the future both fearful and optimistic. When these narratives clash, we are required to rescript our lives, reincorporating the crisis point into a new story about ourselves. Disease, however, forces this rescripting upon us as an urgent imperative:

> In contemporary Western culture, we are encouraged to think of our lives as coherent stories of success, progress and movement. Loss and failure have their place but only as part of a broader picture of ascendence...In the face of a crisis, another story begins and with the power of retrospection the past is rewritten for the exigencies of the present and the future. (1997: 9)

Reisfield and Wilson have noted (2004) that the martial metaphor may be less prevalent today than that of cancer as a 'journey', a metaphor with particular resonance in an era where cancer

> has largely been transformed from an acute event to a chronic illness, enmeshed in life narratives that may span years or even decades. ...The cataclysm of a cancer diagnosis can compel patients to examine the authenticity of their journeys. The exigencies of serious illness can force them to exit the freeway of life on which they had been traveling, often on "cruise control," often at high speed, often with little thought of anything but arriving at the next destination. (2004: 4026)

The substitution of the journey metaphor allows the sufferer, or their families and friends, to integrate the crisis into a new life story, whether as a triumph of strong personality traits and positive thinking, as advocated by the self-help guides Stacey analysed (see also Ehrenreich 2010; Frank 1993), or as a chance to change direction, re-evaluate and emerge as a more 'authentic' self.

Ultimately however the analytical affordances of these approaches are self-limiting. Sontag's vision of illness stripped of its terror when stripped of its associations appears both naïve, insofar as such an anti-hermeneutics goes against the necessity of the kind of self-rescripting described by Stacey (1997), and itself undesirable, creating, even if realisable, a semiotic vacuum in which darker narratives may proliferate. The ultimate extension of this is also problematic, fuelling the belief for some theorists that the narrative framing of illness may prove more compelling than the illness itself. There are likely few medical specialisms that would find grounds to agree with Eisenberg's rather sweeping assertion:

> The decision to seek medical consultation is a request for interpretation.... Patient and doctor together reconstruct the meaning of events in a shared mythopoesis.... Once things fall in place; once experience and interpretation appear to coincide; once the patient has a coherent "explanation" which leaves him no longer feeling the victim of the inexplicable and the uncontrollable, the symptoms are, usually, exorcised. (1981: 245)

Sontag's work has continued to find relevance in clinical and practitioner contexts where the aim is to root out the symbolic bases of treatment failure and patient suffering. These analyses are, rightly, aimed at identifying disfunctions which arise from overloaded or 'pathological' semiosis. They are palliative and goal oriented in focus and ultimately reduce the question of representation down to an instrumental level. How can we, as carers, clinicians, a society as a whole, remove, ameliorate or substitute one set of associations for another we see as more enabling? Whilst the effort is noble, the horizon of the analysis is limited.

SOCIAL REPRESENTATIONS THEORY

An alternative approach to looking at representations of illness comes out of the work of Sergei Moscovici. Largely because of the uptake and popularisation of his work on psychiatry by, in the first instance, his doctoral students from the 1970s (Herzlich 1973; Jodelet 1991), Moscovici's approach has had a wider airing in health studies than in mainstream media and sociological studies. Moscovici (2000) approaches social representations as a rarefication of the concept of collective representations as this appears in the work of Durkheim. For Moscovici, social representations constitute a third path between pure social constructionism/social determinism and a raw intellectual materialism through historicising the process of meaning attribution. Thus, he argues that the 'peculiar power and clarity of representations – that is, of social representation – derives from the success with which they control the reality of today through that of yesterday and the continuity which this presupposes' (2000: 24). Communities continually revivify meaning in the form of 'common sense', or what 'everyone knows' (ibid.: 67) and in so doing create a common base of social solidarity and community cohesion through this knowledge. For Moscovici, this is the central element in the search for an explanation of how signification and meaning creation occur in the first place, the dilemma at the heart of the constructivist/materialist dichotomy referred to above. He argues that communities produce social representations in response to a cognitive need, the affective security of the known. As a species humans are disturbed and troubled less by clear threats than the ambiguous, things which appear to us to cross boundaries (Douglas 1966/2001). This takes the form of an anxiety around the 'not-quite rightness' (ibid.: 38) of something, a Frankenstein which provokes a sense of indefinable dread. We seek to soothe our collective distress by 'transferring what disturbs us…from the outside to the inside…effected by separating normally linked concepts and perceptions and setting them in a context where the unusual becomes usual, where the unknown can be included in an acknowledged category' (ibid.: 39). It is for this reason, Moscovici argues, that representations proliferate and are multiply contested and subject to a flurry of signification at times of societal stress. This argument puts flesh on the bones of the kinds of social strain strand of moral panic theories (Merton 1938; Cohen 1987) in which direct links between rising social tension and the emergence of specific scapegoats is traced.

The keys to understanding representation in Moscovici's work lie in the dual processes of anchoring and objectification. Anchoring is the process of naming and including an object, person or process in a category; their occupation of that category will then inform the attributes we attach to them. Herzlich (1973) uses the example of fatigue here. Many definitions of fatigue exist, and there are multiple ways in which we can experience this. However, once we separate 'fatigue' from other bodily states, we can medicalise it, attribute rights to those who suffer from it normally withheld from those who do not occupy a social 'sick role' (Parsons 1951), and so forth. In anchoring though, we do not as a society spin through a mental Rolodex of possible categories and find the best fit. Rather, we use two contradictory tools, generalising and particularising. Generalising allows us to find a broad category, and particularising notes the distance from the general category occupied by the specific referent. Moscovici uses the example of psychoanalysis to illustrate this point:

> Thus when trying to define and make more accessible the psychoanalyst's dealings with his patient – that 'medical treatment without medicine' which seems eminently paradoxical to our culture – some people will compare it to 'a confession'. The concept is thus detached from its analytical context, and transposed to one of priests and penitents of father confessors, and contrite sinners. Then the method of free association is likened to the rules of confession. In this way, what had first seemed offensive and paradoxical becomes an ordinary, normal process. Psychoanalysis is no more than form for confession. And later, when psychoanalysis has been accepted as a social representation its own right, confession is seen, more or less, as a form of psychoanalysis. Once the method of free association has been separated from its theoretical context and given religious connotations it ceases to be surprising and disturbing and assumes instead a very ordinary character. And this is not, as we might be tempted to believe, a simple matter of analogy but an actual, socially significant merging, a shifting of values and feelings. (2000: 39)

A specific representation is therefore a function of prototype plus qualification, and the process of anchoring is merging two or more phenomena into a single broad category which does not exemplify in the first instance. In the case of the above example, 'confession' and 'free association' come to be anchored together under the category of 'therapeutic talk', which becomes the basic category and is then given a spiritual or medical inflection depending on context. The process of anchoring displaces the original subject, making it a subservient part of a bigger category. Thus, confession becomes a sub-category of that of which it was once the entirety.

The second process is that of objectification, taking an unfamiliar object, concept, person or thing and 'saturating' it with reality.

'Representations…restore collective awareness and give it shape, explaining objects and events so that they become accessible to everyone' (ibid.: 36). The attributes and qualities of a thing become available outside its native category for more general explanation. Thus, again using the example of psychoanalysis, once the alien nature of the treatment becomes normalised and anchored, it becomes a social 'common sense' to see a split between conscious and unconscious, to understand behaviour as driven by 'complexes' and so forth. It becomes then

> a key that opened the locks of private, public and political existence. Its figurative paradigm was detached from its original milieu by continuous use and acquired a sort of independence, just as a well-worn saying is gradually detached from the person who first said it and becomes an unmediated fact. (2000: 50–1)

Moscovici's approach has been particularly valuable in health studies, in large part down to its analytic capacity in accounting for three areas which, if not unique to health, are of particular significance in this area. The first is the way this approach accounts for the emergence of new 'problems', providing a way to capture the moment of social rupture which calls those representations into being. Although we live in a society characterised by continual scientific discovery and technological change, these are rarely 'lived' socially, and intensely, as much as medical and health discoveries and changes are. The diagnosis of new or apparently new illnesses, viruses or conditions (e.g. HIV, SARS, chronic fatigue) and the availability of new methods of treatment or bodily imaging (consider, for example, the impact of ultrasound visualisation on both maternal and foetal medicine and the politics of abortion worldwide) are felt more intensely and more collectively than impactful developments in, for example, space exploration or artificial intelligence, however significant these are. The latter are lived for a general public as spectacle, emotionally invested within their originating communities and their figurative paradigms.

Which brings us to the second point, that social representations theories offer us a way to track changing and alternate meanings across different originating communities without either falling into the theoretical void of pure social construction or treating meaning as an emergent property of social structure. The former, lacking a way of accounting for consistent meaning, falls quickly into an overly voluntaristic accounting; the latter assumes that meaning will follow from social dominance. Neither approach can adequately account for mechanisms of change and transformation in meanings. Moscovici's work enshrines a distinction between what he terms the consensual and reified domains of representation, which allows the question of polysemy to be treated empirically rather than a priori (Howarth 2004; Jovchelovitch 1997). As Jodelet (1991) notes, the approach is fundamentally empirical and observational in nature, focused on semantics in use, tied to a practical social scientific weltanschauung and critical of the theoretical austerities in humanities approaches to representations. As Voelklein and Howarth explain:

> The consensual universe is the world of common sense. This is often seen as the space in which social representations are created, negotiated and transformed. The reified universe, by contrast, is inhabited by 'experts', often seen as scientists, who base their judgements of reality on experimentation, logic and rational choice. Moscovici ... has described ideology as a mediator between these two universes. (2005: 18)

Ideology informs the uses that can be made of the different 'knowledges' of expert (reified) and lay (common sense) universes, and orders and distributes their different applications. Thus, social representations theory has analytical utility in looking at what we might term 'crossing domains', in situations in which the consensual and reified domains become blurred.

The politics of health presents us with an excellent arena for looking at how this plays out in practice. Consider, for, example the ways in which different situated knowledges shift in the practice of medical management of labour and childbirth. Since the 1970s in the UK, some practices, discourses and ideas came to be attributed to a reified universe of roles in which participation is a function of expertise. Knowledge and representations are contested through mutually elaborative systems of interlocking roles (mother, doctor, midwife, doula), and each participant is accorded knowledge only in respect of their role(s), which are unequal in power. Social activism has focused on altering the specifics of these power relations, through the work of lobby groups such as the National Childbirth Trust, professional associations (the Nursing and Midwifery Council), trades unions (Royal College of Nursing), patient advocacy groups (in the UK Patient Advice and Liaison Service), and online and in person parental support groups. These activities contest and reassign meaning to practices, but representations remain tied to the idea of an inherent conflict between groups.

We can draw a distinction between this and the consensual universe in which individuals 'are equal and free, each entitled to speak in the name of the group and under its aegis…In this respect everybody acts as a responsible "amateur" or "curious observer"' (Moscovici 2000: 34). This has similarities to the 'ideal speech situation' described by Habermas but crucially places lay and professional knowledges as something which need to be reconciled through finding common ground—emergent sources of stability—rather than by subjecting claims to the arbitration of reason.

Moscovici's approach is not without its critics (see in particular Potter and Litton 1985; Parker 1987), and especially the perspective's ability to handle a genuinely critical analysis of representations has been brought into question (see Jahoda 1988; Billig 1987). Haworth, whilst a key exponent of social representations theory, identifies three key elements needed to coax out it's latent critical capacity: a clearer focus on the relationship between the psychological and social domains of the theory; an elaboration of the process of reification and legitimisation of different knowledge systems (in short how lay knowledge becomes accepted by experts and vice versa); and a more robust formulation of the nature and extent of agency and resistance in identity construction (2004: 13). 'Conceptualised in this way', argue Callaghan and Augoustinos, 'SRT provides a robust and comprehensive framework for analysing how science is constituted as a rhetorical resource for both sides of the debate' (2013: online). It is particularly useful in application to the issue of pro-vaccination and anti-vaccination discourse insofar as, as Callaghan and Augoustinos (ibid.) also identify, it avoids the necessary elitism inherent in the information—deficit model of science communication, in which the challenge of science in public is that of communicating efficiently and unidirectionally from scientists to the public. Knowledges are not seen as inherently either dominant/subservient or necessarily as in conflict but that an intervening belief system, ideology in action, makes them so.

CRITICAL DISCOURSE ANALYSIS

Which brings us on to the final approach to be considered here, namely critical discourse analysis (CDA). This approach emanates from a diverse set on influences in critical linguistics (see Kress and Van Leeuwen 1996; O'Halloran 2007) and uses analyses of the social relations of speech and text to account for the way meaning is conferred. CDA aims to capture not merely the 'how' of language, the ways in which signification is achieved, but further to link language, power and ideology in a way that allows the reader to observe structural inequalities as they are revealed in representations, and to account for the ways in which they are produced and maintained. Where critical linguistics on the one hand is tied to a descriptive engagement with a corpus of material, CDA aims to link this up with an account of behaviours, goals, values and participants and to firmly place the interpreting subject/reader at the heart of the analysis.

For CDA, the 'imagined' subject/reader is the essential anchor insofar as they are themselves outcomes of the process of discourse formation, and thereafter go on to become the way to secure different meanings within it. In this way CDA is an extension of the central tenet of Althusser that ideology is less an external force which obliges compliance but a force which shapes a subject through 'interpellation' or 'hailing' them, addressing them as subjects already compliant. Thus, for example, De Cilla et al. (1999) in this tradition developed their discourse-historical method to account for and to capture the ways in which, '(t)hrough discourse, social actors constitute knowledge, situations, social roles as well as identities and interpersonal relations between various interacting social groups' (1999: 157). Interpellated individuals inhabit, speak from and interpret their subject positions according to the overall distribution of 'subjectivities' available. These are contextually, situationally specific and therefore disaggregated. Individuals are comprised of multiple, sometimes complementary and sometimes competing subject positions within discourses, and the 'self' in this approach refers to the co-ordination of these. A specific subject position directs both understanding and action, narrowing the available interpretive range of a given sign or practice. It is, then, a way to account for polysemy without falling into an unproductive materialist/constructionist 'black hole', as discussed above, insofar as the predictability of the reading, the attribution of meaning, is secured on the reader, not a simple transfer from the text. The reader however is not figured as one who has carte blanche to interpret according to whim, or from the point of view of all possible readings, but only from the point of view of those readings which are available to them given their position within the discourse as a whole.

CDA uses a variety of foci to excavate the linguistic and social underpinnings of discourse. In the first, case studies will focus on the selection of parallel descriptors as a way to access the structure of the 'field' as a whole, a means to capture the ways in which lexical choices signal and invoke broader discourses by seeing how they map onto the overall distribution of alternative terminology. Thus, for example, Van Dijk (1993) demonstrates the ways in which different ways of describing race and ethnicity co-vary with competing attributions of agency and power.

A second concern is with the modalities of discourse or the ways that lexical choices reveal attachment to specific positions whilst evidencing distance from others. Just as specific lexical choices reveal the overall 'pool' or map of alternative terminology associated with different subject positions, so modalities, classifications and modes of persuasion reveal the broad landscape of discourse around a topic, group or idea. To take an example from anti-vaccination discourse, the subject position attached to parents who are questioning medical advice could encompass 'justice seeker', 'insider informant', 'anxious parent', or 'crank', 'troublemaker' or 'hysteric'. In an example from *The Daily Mail* (see below, p.), we can see how the use of anecdote as evidence and the choice of colloquial terminology 'shuffled out', 'bee in his bonnet' and critical imagery ('conveyorbelt', for example) simultaneously acknowledge/describe the field and disavow the speaker from the more discreditable positions within it.

Additionally, although not initially central to the approach, CDA is able to make good use of the analysis of metaphor. Hart (2008) and later Musolff (2012) outlined a way for metaphor to be incorporated in CDA, which avoided importing the seemingly necessary contradictions which had been encountered in conceptual metaphor theory (CMT). CMT and CDA conflict around a central question, the arbitrariness or degree of flexibility available in the selection of 'source domains'. Lakoff and Johnson (1980) divide a metaphor into two elements, the 'source' and the 'target' domains. A 'source' domain is an already-known concept which we use to organise perception of the unknown or contested, in this case the target domain. This is the mechanism by which we come to understand one thing by means of another, and the means by which meaning comes to be attached to a thing. As Lakoff and Johnson point out, these are inherently heterogenous and must be so for a metaphor to work as such. No closely related lexical substitution can function as a metaphor. So far so good. However, for CMT source domains are generated by embodied experience and have as such a material basis and degree of determination. For CDA, on the other hand, the selection of a source domain is a purposive social act which reveals the operation of ideology. So, for example, if we consider this extract from *The Daily Telegraph* from May 1955, under the headline *Polio Vaccine 'Short Cut' Attempted*:

> Mr. Eisenhower conceded today that the medical experts in charge of the Salk polio vaccine programme may have yielded too quickly to strong public pressure for its immediate use. They may, as he put it, have taken a "short cut". The President made the statement at his Press conference when asked why the retesting programme now being conducted by the public health service had not been conducted in advance of any inoculations. He reaffirmed his faith in the efficacy of the vaccine, saying that he had absolute reliance on his medical advisers. ('Polio Vaccine "Short Cut" Attempted', *Daily Telegraph*, 12 May 1955, p. 12.)

Here, the metaphor of a 'short cut' offers a more value (and blame) -laden way to express the concept of speeding up the programme than alternatives might. A shortcut comes to us from the source domain of travel, a journey, but one of a specific nature, one which also implies a reduction of an experience. If we take a shortcut, we don't follow the path, we go 'off map', and Western narratives are replete with examples of the risks of this tendency. So, the target domain of 'rapid delivery of a vaccine' becomes wrapped up with a metaphor which connotes risks to the unwary or of negligence. If of course we altered the descriptor whilst retaining the source domain of a journey (e.g. 'fast track' and 'wrong turn'), we would have a different set of connotations all together. Alternatively, if we substitute the source domain, this would again differ. For example, if we were to use the source domain of sport rather than journey (e.g. 'they may have jumped the gun', or 'fumbled the pass'), the metaphor would carry connotations of well-intentioned failure rather than of negligence. The choice of metaphor is therefore never innocent, and it is persuasive insofar as it is an abstraction, implying a common essence between two disparate phenomena. If we tried to convert the metaphor into a statement (e.g. scientists moved too quickly/ did not follow procedures), it could be answered empirically, confirmed or refuted, and the truth of the statement would be synthetic, observable in the fit with real-world practice. At the level of abstraction, of metaphor, its 'truth' is analytic in nature, that those who take shortcuts, procedural or geographical, come to the same common, sticky end.

CDA is capable of encompassing all aspects of representation through language, or as in the case of multi-modal critical discourse analysis

(MCDA) visual communication. Which aspect of language carries the working 'load' of representation within a specific discourse will naturally vary along with the nature of the discourse itself. So, for example, looking at linguistic modalities, here considered as how a speaker signals distance or proximity to a statement, can be more fruitful in some circumstances than analysing metaphor. This is certainly the case in analysing scientific discourse as this is manifested within popular culture and communication. A very good example to illustrate this is the political mobilisation of proximity and agreement with science in American politics, a discourse organised and inflected by the presence of a populist anti-intellectualism which some have described as a 'war on science'. A moment of discursive 'rupture' in this was the use/overuse of the phrase 'I'm not a scientist' by Republican candidates in the run up to the 2016 presidential election. The phrase was originally a strategy used to deflect criticism and avoid discussion in areas where their policies conflicted with environmental concerns. However, what was first an intellectual mea culpa quickly became connotatively loaded, a linguistic shortcut to signal Republican identification with 'ordinary folks' and a badge of anti-intellectual pride. 'I'm not a scientist' became a shortcut signalling that the speaker was a freethinker. Democrat President Barak Obama made a response to this a key part of his 2015 State of the Union speech:

> I've heard some folks try to dodge the evidence by saying they're not scientists; that we don't have enough information to act. Well, I'm not a scientist, either. But you know what—I know a lot of really good scientists at NASA, and NOAA, and at our major universities. The best scientists in the world are all telling us that our activities are changing the climate, and if we do not act forcefully, we'll continue to see rising oceans, longer, hotter heat waves, dangerous droughts and floods, and massive disruptions that can trigger greater migration, conflict, and hunger around the globe. The Pentagon says that climate change poses immediate risks to our national security. We should act like it.

The modalities of this are complex. In the use of 'best scientists in the world are all telling us', the 'Pentagon says that', 'really good scientists at NASA, and NOAA, and at our major universities', Obama expresses alignment with scientific insights and, as significantly, brands them as American attributes (*our* universities, for example). But also significant is the way he presents his own knowledge in respect of this. 'I've heard' and 'But you

know what—I know a lot of really good scientists' locate Obama's expertise as observation and mediation and invites the listener to enjoy the same
position (you know/I know). Thus, in this speech alignment with science
is not an unthinking capitulation to expertise of the 'scientists know best'
variety but an acceptance of contrasting but complementary areas of
expertise and authority. These two modalities are, of course, mutually co-
organising. In this respect, recognising the modalities in this speech gives
us a richer understanding of the discourse around science and politics than
a metaphorical analysis would offer. Similar points can be made concerning other formal techniques used in texts, for example the suppression of
implied terminology, personification, objectification, metonymy and syn-
ecdoche. Each element can carry greater weight in an analysis according to
the discourse itself, and each is seen in CDA as a purposive and motivated
choice by the speaker.

In terms of this analysis, it is CDA's capacity to account for meaning
through modality that makes it invaluable. One of the key problems in the
use of metaphor analysis in historical work is the extent to which the pool
of available substitutions can or cannot be reconstructed. Whether we
understand metaphors, as in CMT, as emanating from the experiential, the
embodied encounter with the world, or, as in CDA, as more arbitrarily
available but ideologically directed, the fact remains that they can only act
to port meaning from one stable domain to another if they are, as Lakoff
and Johnson (1980) assert, different from each other. This requires us to
understand ahead of our analysis how the potential source domains are
themselves organised within discourse at the period that we are looking at.
So, to return to the example of the 'short cut', our comprehension of how
this worked in 1955 is predicated on the knowledge that science is understood as 'progress', that to progress is a metaphor of movement, and so
on. It is also predicated on our unconscious sense of what movement
involves. The metaphor of a shortcut has little resonance in a wide-open
landscape, more in an urban one. To pardon a pun, we could tie ourselves
in the most dreadful of knots if we sought after the moment in the past at
which the knot became available to our experience as something which
could act as a metaphor. Without this sense of what other substitutions are
available then, our ability to be confident in the ideological motivation for
the selection of one or another is weak.

Representations from Health and Illness to Vaccination: The Stories We Tell

The final streams I'm going to look at here are representations of epidemics on the one hand and vaccinations on the other in popular culture and the ways in which academics have theorised these. One key approach comes from analyses of 'urban legends' and narratives; we should start with the work of Priscilla Wald on 'outbreak' narratives. Wald argues that societies produce a formulaic narratives or plots in the face of new epidemics of infection. These act as 'an explanatory story that is not specifically authored, but emerges from a group as an expression of the origins and terms of its collective identity. Its strong emotional appeal derives from and affirms the fundamental values, hierarchies, and taxonomies that are the preconditions of that identity' (Wald 2008: 9). In the case of outbreaks, this narrative is heavily focused on networks of transmission, and

> chronicles the epidemiological work that ends with its containment. As epidemiologists trace the routes of the microbes, they catalog the spaces and interactions of global modernity. Microbes, spaces, and interactions blend together as they animate the landscape and motivate the plot of the outbreak narrative: a contradictory but compelling story of the perils of human interdependence and the triumph of human connection and cooperation, scientific authority and the evolutionary advantages of the microbe, ecological balance and impending disaster. (2008: 2)

In so doing, outbreak narratives have, even more so than other social myths, an inherent bias towards the reassertion of community and its boundaries. They 'promote or mitigate the stigmatizing of individuals, groups, populations, locales (regional and global), behaviors, and lifestyles' (2008: 3).

Howsoever tempting it may be to ascribe this universality to the outbreak myth, it is hard to sustain some of its premises. The recent pandemic certainly witnessed some of the global tracing that Wald describes, but whereas in 2008 this acted to reassert global interdependence, Covid represented this in more threatening and competitive terms. The reliance of nations and governments on digital 'track and trace' technologies posed the problem of exposure and infection on an individual level, giving rise to a culture of blame within communities. Although the scapegoating of some communities was entirely consistent with the outbreak narrative, national and localised lockdowns and shelter-in-place orders redefined micro-communities and rarefied dichotomies of 'us' versus 'them'. Also noticeably absent in the Covid-19 narratives was the 'resolution' phase identified by Wald in which the 'mythic features of the narrative, in turn, temper the urgency with an implicit promise not only of survival but also of renewal. Even as the species war threatens apocalypse, the conventionality of the story anticipates the triumph of science and epidemiology and affirms the worth of humanity' (2008: 268). In the UK, at least such affirmations were transcribed onto individuals or, as in the case of the weekly 'clap for the NHS', professions whose roles were predominantly sacrificial. The anticipation of triumph was largely absent.

Of course, Wald's work focused on the immediate post-SARS period, a point at which internationalism and globalisation were largely taken for granted within popular and political discourse. Covid, as I will be arguing below, took place against the background, already well in process, of the collapsing of supranational institutions and bodies. The ongoing withdrawal of the US from international regulatory organisations, the rise of right-wing populism in the West, the loss of cosmopolitan ambition on the part of the UK and what some have described as the increasing fortification of Europe are all features of this. As Schama has noted of Covid-19:

> You would suppose that in the face of a pandemic, an outbreak, which by definition is global together with a recognition of shared vulnerability, governments and politicians might have set aside the usual mutual suspicions, and under the aegis of the WHO, agreed on common approaches to containment, vaccination and control. Needless to say, nothing remotely like that has happened. If anything, the reverse has been the case: responses to the pandemic sharply diverged, even within entities like the European Union, ostensibly committed to common policies. Decisions taken by individual American states on vaccination requirements and mask mandates thwarted federal guidelines deepening the already bitter cultural divisions between 'red' and 'blue' America. (2023: 31–2)

This is all the more ironic, he goes on to observe, as, according to WHO Director General Tedros Adhenom Ghebreyesus, 'until everyone is vaccinated, no one will be safe' (ibid.: 2023).

There is, as many have observed, an overlap between the world of urban myths and that of conspiracy theories insofar as stories that we would see as part of an ad hoc mythology are often appropriated into the discourse of conspiracy theories. However, although many scholars have had recourse to conspiracy theories in describing the communicative forms and content of anti-vaccination materials, there is often a lack of a clear sense of where other genres of communication end and conspiracy theories begin. Such accounts most often depend on a 'commonsense' approach to conspiracy theories, which, as Baden and Sharon (2021) point out, is likely to devolve down to little more than an apprehension that an alternative position is based on a dogmatic defense of 'faulty knowledge'. The authors offer a useful alternative model to identify conspiracy theories, proposing

> that CTs (conspiracy theories) that deserve to be disqualified as violations of reasonable democratic discourse—can be delineated as theories that invoke grand conspiracies to account for social phenomena and additionally meet three distinguishing criteria: They assume: (a) conspirators' unrealistically encompassing knowledge and control over events (pervasive potency), b) construct an essentialist binary of truth and falsehood, good and evil (Manichean binary), and c) systematically debase conventional institutions and processes of epistemological validation (elusive epistemology). (2021: 90)

This offers us a rather clearer path, one which has less risk of marginalising and dismissing other knowledges.

The common set of representational strategies which anti-vaccination discourse is seen to derive from conspiracy theories are, firstly, a set of binaries around social actors. From the point of view of anti-vaccination 'believers', this binary is around 'responsible parent' versus 'interfering authority'. For those who advocate immunisation, the same binary structure is operative, this time between 'dogmatic/ ignorant' versus 'knowledge giver'. These binaries replicate within other aspects of anti-vaccination and pro-vaccination discourse, for example 'big pharma/home remedy', 'pollution/purity' and so forth. Secondly, the theme of pervasive potency is also attributed to anti-vaccination discourse. In the case of Covid-19, for example, we saw this playing out through the media coverage of the Wuhan Laboratory and in the belief that this was the source of the outbreak. The search for a single point of origin is though also a feature of other narratives—for example, the 'outbreak' narrative identified by Wald (see above) and not specific to conspiracy theories as such. Pervasive potency is also figured in media coverage of polio and pertussis, as will be discussed in Chaps. 4 and 5 respectively. The final point made by Baden and Sharon concerning epistemological validation and its refutation is a little complex here. As observed above, in respect of Moscovici, the work of representation and the ways in which new meanings come to be attached to objects and practices is tied up with the transfer of interpretations from one 'universe' to 'another', the 'reified' to the 'consensual'. It is a work of subtlety to adequately separate the 'elusive epistemology' of conspiracy theories from the normal resistance attending the transfer from one domain to another. This is similar to the point raised by Barkun (2015) in his work on conspiracy theories as 'stigmatized knowledge'. For Barkun, the common attributions made concerning conspiracy theories, that they are eclectic assemblages, insofar as a belief in one leads to a requirement for acceptance of all and that the main test of truth for conspiracy theories is they are rejected by 'authorities', are both becoming weaker. He argues that marginalised knowledge is no longer as clearly differentiated from mainstream knowledge and that the gap is continuing to narrow, in part as a result of social media and internet technologies. The process by which these boundaries break down, Barkun argues, is tied to the transfer of stigmatised knowledges from their communities of origin to a wider public as they come to be represented, positively or negatively, in mainstream and especially entertainment media. Citing examples from film to novels, Barkun argues that the content of conspiracy theories has become 'mainstreamed' with 'ever-increasing amounts of what once would have been considered fringe motifs … finding their way into channels that reach mass audiences' (Barkun 2015: 117). Once more widely available, marginalised knowledges, disassociated from their discredited points of origin, can become compelling, especially to the extent that either they become ravelled up with other previously stigmatised knowledges/issues which themselves come to be more credible (a likely outcome where conspiracy theories tend to cluster together) or through a form of 'confirmation bias', in which the mere appearance of claims in multiple spheres stands proxy for truth.

In terms of representations within anti-vaccination discourse specifically, separate from conspiracy theories, there are relatively few studies available on which to draw (see Hobson-West 2007). DiRussso and Stansberry are a notable exception to this. In their study of online communities of modern 'anti-vaxxers', they identified a range of recurrent key themes which structure anti-vaccination discourse. Firstly, there is the binary noted above between artificial 'pharma' and the 'healthy lifestyle' which promotes 'natural immunity', a framing which accords with a wider emphasis on 'God's law' or the 'natural order' as an alternative to medical and scientific arguments. There was also a clear and understandable bias towards explanations which complemented and reiterated core communal values. Thus, the authors noted that parents were unwilling to discuss the HPV vaccine with their (adolescent) children, likely tied to a wider commitment around discussion of sexual behaviour and sexuality. We might also note in this context the prevalence of explanations of anti-vaccination stances which are based upon a valorisation of traditional family structures and the autonomy of parents in making their own decisions about families. DiRussso and Stansberry also noted the extent to which the 'anti-vaxxers' they studied resisted the label of 'anti-vaxxer', styling themselves instead as 'enlightened, not opposed'. 'Anti-vaccination materials often feature explicit statements declaring those who question vaccine safety as more informed and aware than the general population…advocates urge pro-vaccination responders to "get educated" and "learn the truth" about vaccines. Anti-vaccination messaging frequently refers to the general population as "sheep" who need to "wake up"' (DiRusso and Stansberry 2022: 323). The sheep/'sheeple' motif was, incidentally, one which featured heavily in anti-vaccination discourse over the course of the Covid pandemic.

These findings have continuities with those of Hobson-West (2007), who looked at emerging 'Vaccine Critical' groups in the UK. Hobson-West identifies three key senses in which their discourse differs from that of prior anti-vaccination communications. On the one hand, she argues there is a rarefication of the category of the good parent such that 'the good parent becomes one who spends the time to become informed and educated about vaccination' (2007). Secondly, Vaccine Critical groups 'construct trust in others as passive and the easy option. Rather than trust in experts, the alternative scenario is of a parent who becomes the expert themselves, through a difficult process of personal education and empowerment. Note that this discourse demands a process of education. Trust in self is not assumed to be automatic or pre-existing'. Finally, she notes, this new emphasis places vaccine criticalness on a continuum with other kinds of patient advocacy, including the (positively regarded) model of the patient-expert/health consumer (2007: 212). Interestingly, none of these categories fit well with the conspiracy theory narrative/genre.

This chapter has reviewed a number of key themes and approaches to develop a 'toolkit' for digging into representations of vaccines, immunisation and the vaccine hesitant. These tools are drawn from Sontag's work on the representational politics of health; social representations theory's concern with the communities construed by the processes of anchoring and infusing new objects, groups, categories and experiences with extant meaning; and CDA's focus on the way we can apply modalities of discourse to the task of contentious representations. We have also considered how more specific representational 'universes', that of the 'outbreak narrative' and that of 'conspiracy theories', shape the forms of anti-vaccination discourse from outside. We will now go on to look at the tools available on the other side of the equation, quantitative and relational analyses.

REFERENCES

Baden, C. and Sharon, T. (2021) 'BLINDED BY THE LIES? Toward an integrated definition of conspiracy theories', Communication Theory, 31(1), pp. 82–106. Available at: https://doi.org/10.1093/ct/qtaa023.

Barkun, M. (2015) 'Conspiracy theories as stigmatized knowledge', Diogenes, 62(3–4), pp. 114–120. Available at: https://doi.org/10.1177/0392192116669288.

Billig, M. (1987) Arguing and thinking. A rhetorical approach to social psychology. Cambridge: Cambridge University Press.

Brandt, A.M. and Botelho, A. (2020) 'Not a Perfect Storm – Covid-19 and the Importance of Language', New England Journal of Medicine, 382(16), pp. 1493–1495. Available at: https://doi.org/10.1056/NEJMp2005032.

Callaghan, P. and Augoustinos, M. (2013) 'Reified versus consensual knowledge as rhetorical resources for debating climate change', Revue internationale de psychologie sociale, 26(3), pp. 11–38.

Chopra, A. K., & Doody, G. A. (2007). Schizophrenia, an illness and a metaphor: analysis of the use of the term 'schizophrenia' in the UK national newspapers. Journal of the Royal Society of Medicine, 100(9), 423–426.

Clow, B. (2001) 'Who's Afraid of Susan Sontag? or, the Myths and Metaphors of Cancer Reconsidered', Social History of Medicine, 14(2), pp. 293–312. Available at: https://doi.org/10.1093/shm/14.2.293.

Cohen, Stanley. (1987) Folk devils & moral panics: the creation of the Mods and Rockers. [Third edition]. Oxford: Basil Blackwell.

Coulehan, J. (2003) "Metaphor and medicine: narrative in clinical practice." The Yale Journal of Biology and Medicine 76 (2003): 87–95.

Craig, D. (2020) 'Pandemic and its metaphors: Sontag revisited in the COVID-19 era', European Journal of Cultural Studies, 23(6), pp. 1025–1032.

Davis, C.J. (2002) 'Contagion as Metaphor', American Literary History, 14(4), pp. 828–836.

De Cillia, R., Reisigl, M. and Wodak, R. (1999) 'The discursive construction of national identities', Discourse & Society, 10(2), pp. 149–173.

DiRusso, C. and Stansberry, K. (2022) 'Unvaxxed: A Cultural Study of the Online Anti-Vaccination Movement', Qualitative Health Research, 32(2), pp. 317–329. Available at: https://doi.org/10.1177/10497323211056050.

Douglas, M. (1966) Purity and Danger: AN ANALYSIS OF THE CONCEPTS OF POLLUTION AND TABOO. London and New York: Routledge.

Ehrenreich, B. (2010) Smile or Die: How Positive Thinking Fooled America and the World. London: Granta.

Eisenberg, L. (1981) 'The physician as interpreter: Ascribing meaning to the illness experience.', Comprehensive Psychiatry, 22, pp. 239–248.

Fasce, A. et al. (2023) 'A taxonomy of anti-vaccination arguments from a systematic literature review and text modelling.', Nature human behaviour, 7(9), pp. 1462–1480.

Frank, A.W. (1993) 'The rhetoric of self-change: Illness experience as narrative', The Sociological Quarterly, 34(1), pp. 39–52.

van der Geest, S. and Whyte, S.R. (1989) 'The Charm of Medicines: Metaphors and Metonyms', Medical Anthropology Quarterly, 3(4), pp. 345–367. Available at: https://doi.org/10.1525/maq.1989.3.4.02a00030.

Hanne, M., Hawken, S. and Sontag, S. (2007) 'Metaphors for illness in contemporary media.' Medical Humanities, 33, pp. 93–99.

Hart, C. (2008) 'Critical discourse analysis and metaphor: Toward a theoretical framework.', Critical Discourse Studies, 5(2), pp. 91–106.

Herzlich, C. (1973) Health and Illness: A Social Psychological Analysis. London: Academic Press.

Herzlich, C. and Pierret, J. (1987) Illness and Self in Society. Baltimore: Johns Hopkins University Press.

Hobson-West, P. (2007) '"Trusting blindly can be the biggest risk of all": organised resistance to childhood vaccination in the UK', Sociology of Health & Illness, 29(2), pp. 198–215. Available at: https://doi.org/10.1111/j.1467-9566.2007.00544.x.

Howarth, C. (2004) 'Re-presentation and resistance in the context of school exclusion: Reasons to be critical', Journal of Community and Applied Social Psychology, 14, pp. 356–377.

Jahoda, G. (1988) 'Critical notes and reflections on "social representations"', European Journal of Social Psychology, 18(3), pp. 195–209. Available at: https://doi.org/10.1002/ejsp.2420180302.

Jodelet, D. (1991) Madness and Social Representations: Living with the Mad in One French Community. New Jersey: Prentice Hall.

Jovchelovitch, S. (1997) 'Peripheral communities and the transformation of social representations: Queries on power and recognition.', Social Psychological Review, 1(1), pp. 16–26.

Kleinman, A. (2020) Illness Narratives : Suffering, Healing, and the Human Condition. New York: Basic Books.

Kress, G. and Van Leeuwen, T. (1996) Reading Images: The Grammar of Visual Design. 2nd edn. London: Routledge.

Lakoff, George. and Johnson, M. (1980) Metaphors we live by. Chicago ; University of Chicago Press.

Merton, R.K. (1938) 'Social Structure and Anomie', American sociological review, 3(5), p. pp. 672–682.

Moscovici, S. (2000) Social Representations: Explorations in Social Psychology. Edited by G. Duveen. Cambridge: Polity.

Musolff, A. (2012) 'The study of metaphor as part of critical discourse analysis', Critical Discourse Studies, 9(3), pp. 301–310. Available at: https://doi.org/1 0.1080/17405904.2012.688300.

Nuwarda RF, Ramzan I, Weekes L, Kayser V. (2022) Vaccine Hesitancy: Contemporary Issues and Historical Background. Vaccines (Basel). Sep 22;10(10):1595. doi: https://doi.org/10.3390/vaccines10101595.

O'Halloran, K. (2007) 'Critical Discourse Analysis and the Corpus-informed Interpretation of Metaphor at the Register Level', Applied Linguistics, 28(1), pp. 1–24. Available at: https://doi.org/10.1093/applin/aml046.

Parker, I. (1987) '"Social representations": Social psychology's (mis)use of sociology', Journal for the Theory of Social Behaviour, 17(4), pp. 447–469. Available at: https://doi.org/10.1111/j.1468-5914.1987.tb00108.x.

Parsons, T. (1951) The social system. London: Routledge & Kegan Paul.

Patterson, J.T. (1987) The dread disease: cancer and modern American culture. Cambridge, MA, Harvard University Press

Potter, J. and Litton, I. (1985) 'Some problems underlying the theory of social representations.' British Journal of Social Psychology, 24, pp. 81–90.

Potts, A. and Semino, E. (2019) 'Cancer as a Metaphor', Metaphor and Symbol, 34(2), pp. 81–95. Available at: https://doi.org/10.1080/1092648 8.2019.1611723.

Reisfield, G.M. and Wilson, G.R. (2004) 'Use of Metaphor in the Discourse on Cancer', Journal of Clinical Oncology, 22(19), pp. 4024–4027. Available at: https://doi.org/10.1200/JCO.2004.03.136.

Rothman, S. M. (1995) Living in the Shadow of Death: Tuberculosis and the Social Experience of Illness in American History Baltimore: Johns Hopkins University Press

Schama, S. (2023) Foreign Bodies: Pandemics, Vaccines and the Health of Nations. New York: Simon & Schuster (epub.).

Small, N., Downs, M. and Froggatt, K. (2006) 'Improving end-of-life care for people with dementia – the benefits of combining UK approaches to palliative care and dementia care', in B. Miesen and G. Jones (eds) Care Giving in Dementia Research and applications. London: Routledge, pp. 365–392.

Sontag, S. (1977) Illness as Metaphor. New York: Farrar, Straus and Giroux.

Sontag, S. (1989) Aids and its metaphors. New York: Farrar, Straus and Giroux.

Stacey (1997) Teratologies: A Cultural Study of Cancer. London: Routledge.

Trčková, D. (2015) 'Representations of Ebola and its victims in liberal American newspapers', Topics in Linguistics, 16(1), pp. 29–41. Available at: https://doi.org/10.2478/topling-2015-0009.

Van Dijk, T. (1993) Discourse and Elite Racism. London: Sage.

Voelklein, C. and Howarth, C. (2005) 'A review of controversies about social representations theory: a British debate', Culture and Psychology, 11(4), pp. 431–454.

Wald, P (2008) Contagious Cultures, Carriers, and the Outbreak Narrative Durham, NC US: Duke University Press.

Wallis, P. and Nerlich, B. "Disease metaphors in new epidemics: the UK media framing of the 2003 SARS epidemic." Social Science & Medicine (1982) 60 (2005): 2629–2639.

Williams Camus, J.T. (2009) "Metaphors of cancer in scientific popularization articles in the British press." Discourse Studies 11 (2009): 465–495.

CHAPTER 3

Relational and 'Big Data' Approaches to Representation in Understanding Illness

Abstract This chapter traces a path for the reader through recent historical analyses which use large-scale data sets. The chapter offers a review of studies which have been influenced by 'big data' approaches in historical contexts and considers the impact of the availability of digitalised archives and digitalisation on the framing of historical research. The chapter outlines distinctions between big data approaches and those using relational data analysis. The work concluded with a consideration of the usefulness of large-scale data approaches in historical periodical archives, arguing that the project of periodical digitalisation presents the potential for biases in pattern recognition.

Keywords Big data • History of newspapers • Apophenia • Digitalisation • Power • Network analysis

This chapter takes an alternative route to looking at how we can assess change and development in anti-vaccination narratives through looking at the affordances of large-scale relational data analysis and what has come to be termed 'big data' approaches. The chapter opens with an introduction to and examination of the impact of big data on historical studies, especially historical studies of the media, before going on to look at how this approach has already been used in the context of attempts to unpack

© The Author(s), under exclusive license to Springer Nature Switzerland AG 2024
A. Cavanagh, *Anti-Vaccination and the Media*,
https://doi.org/10.1007/978-3-031-70559-5_3

representations of health and illness. Finally, I will conclude with a consideration of studies which have already deployed this approach in looking at anti-vaccination narratives and offer some conclusions.

This piece is not the first attempt to reflect on the impact of the availability of large-scale data sets within historical studies of the media. Media history, more so than many other subject areas, has had good cause to take a critical view of these developments as it has been most immediately affected by large-scale digitalisation of archives. In the wake of digitalisation projects, there have been a number of insightful and nuanced considerations of these sorts of impacts (e.g. Bingham 2010; Leary 2005; Nicholson 2013; DiCenzo 2015) but they have usually been confined to looking at this impact within a sub-sphere, in these cases digitalisation of periodicals. Whilst this is valuable, it doesn't give us a guideline for how far we might be able to generalise from the experience of one set of scholars to the wider field of media history. The challenges of using large sets of digitalised materials in different areas may overlap only to some extent. Nor, on the other hand is this the first piece to reflect on the challenges of using big data approaches (BDA for short) in historical studies more generally. The work of Graham et al. (2016); Mayer-Schönberger and Cukier (2013); Tanaka (2013); Madsen-Brooks (2013); and Erickson (2013) are all excellent guides to the uses and pitfalls of big data as applied to heterogenous contexts.

However, these reflections are often dispersed and take as their point of departure either sets of concerns which emerge from particular disciplinary or topic-specific considerations or arise from the use of specific technologies of analysis and data presentation. Such dispersion and fragmentation make it hard to grasp an overview, to see the affordances of these disparate approaches and also to assess the cost of their use. Here, what I want to do is to set big data approaches in context and specifically to draw out some of the implicit but under-explored aspects of the move to this kind of analysis. In particular here I am concerned with four core areas, namely the extent to which big data analysis is driven by a *totalising logic*, which plays out both philosophically and methodologically; the assumptions which underpin a shift from identity to association as a logic of study; the assumptions about the nature and quality of evidence which are required by the use of these approaches; and finally the ways in which a shift to the use of big data reframes, dislocates and relocates core questions in the field of media history. Finally, I argue the way big data has been treated constitutes a 'field generating' move which creates greater specialisation and fragmentation.

Big Data and a Totalising Logic

Big data is often figured as a method of analysis whose time came with the rise of the technology necessary to achieve it, the assumption being that technology merely facilitates what the science required all along. Such an approach, in addition to its teleology, understates the broader cultural roots of this shift towards big data. One cultural shift of recent times, within and without the academy, has been the embrace of a desire to see the social whole, but one which has become dis-embedded from its philosophical and politically critical roots. The desire for a broader perspective maps out in popular culture in a well-documented shift towards the macroscopic in drama and art. Toscano (2012) uses the example of the popularity of dramas like *The Wire* as an instance of this, though the tendency is much broader. The desire to see totality is evidenced in the prevalence of conspiracy theory, here as Jameson's 'poor person's cognitive mapping in the postmodern age' (cited in Aupers 2012: 23); in the popularity of the social panorama; and in the ever-accelerating rise of reality broadcasting and social media narrowcasting. This kind of perspective, Toscano argues, is frustrated by the academy, in which impact is 'channelled into governmental and corporate channels which have little interest in the disquieting, antagonistic or counter-intuitive consequences of trying to think totality' (2012: 65). For Toscano, the academy is lamed by its instrumentalism and seeks to reject totalising perspectives, substituting an atheoretical and therefore apolitical empiricism in its place.

I will be arguing below that, on the contrary, BDA constitutes an academic manifestation of this 'will to totality' but one which is separated from the latter's critical roots and takes on the guise of an empiricism which is anything but apolitical. The move towards BDA is not an extension of simple empiricism so much as an attempt to discover the subject posited in critical theory through empirical means and one which imports into BDA politically motivated assumptions. The will to totality, I argue, can be seen on multiple levels and plays out in combining and importing different types of data and sources. BDA works by bringing together multiple diverse and disparate forms of evidence and subjecting them to the same analytical operations. This removal of data from their contexts changes the nature of both data and our encounters with it. This can be experienced on multiple levels, prosaic and more abstract.

On a purely practical level, we encounter this in everyday decisions on the practical handling of data. Bringing together diversely formatted data lays bare the problems of establishing equivalence between data points. Wright Kennedy et al. (2017), for example, reflect on the inaccessibility of handwritten records (most often the format of historical public records) to modern optical character recognition (OCR). Gupta, Jacobson and Garcia (cited in Graham et al. (2016: 49)) report similar difficulties in using OCR on digitalised newspapers, where variability in fonts, printer noise, line-break hyphenation and unequal spacing in records reduces the effectiveness of the OCR. The tendency is to treat these OCR failures as a technical problem which advances in digitalisation and which OCR should overcome. However, that assumption is valid only to the extent that, firstly, we can be confident that there are stable distinctions in the data between 'signal' and 'noise', what is to be extracted versus what is overcome, and, secondly, that our technical means of rendering the disparate accessible doesn't layer on a false equivalence between data points. As Sjøvaag and Karlsson (2017) point out in discussing links in digital content, the fact of a link doesn't necessarily indicate human agency and therefore cannot be taken as indicating intention. Where links are taken as an index of 'engagement', then, the data points are not equivalent, but after aggregation the means by which these data are generated is occluded. As Kuhn once observed, 'In history, more than in any other discipline I know, the finished product of research disguises the nature of the work that produced it' (cited Tanaka 2013: 44).

On a broader level, BDA dis-embeds data from its ecologies of production and therefore from means of interpretation. Brügger and Finnemann (2013), for example, flag up philosophical and practical problems attendant on relocation of data in their discussion of online digital archives. For them, there are important distinctions between web sources that are analysed in their own environments, 'in the wild' as it were, and those accessed through web archives. In the case of archived materials, they point out, transformations have occurred which are the consequences of selections by archivists. '(W)hat is archived is almost never a 1:1 scale of what was once online' (2013: 74).

Whilst this may seem a variant of a universal problem in historical scholarship, the partial nature of sources, Brügger and Finneman's point is wider. It is not merely that digital materials that are archived are partial selections from a pool of potential choices but more that they are *products* of decisions made by others about their nature. Hyperlinks, for example, can link sections of texts, or whole texts. What a hyperlink refers to though can't be resolved simply by following it. Its logic, whether it is a purposive, goal-oriented link or an associative link, where and when and in response to what it was added, is not bundled with its direction and object. The archive freezes and ossifies web materials and cannot port these meanings alongside them. The process of archiving also collapses together different types of metadata, those that are part of the 'digital born' and those imposed from the outside by archivists. While digital-born and digitised materials may both have similar functionality, as both can be searchable, both can be hypertextual, the authors point out the form of these functions are 'particular for each medium' (ibid.: 70). For example, '(w)hile a free text search leaves a wide semantic variation open to the user, a meta-tag-based search limits the variation in favor of more goal-oriented link relations' (ibid.: 71). In this case, the affordances of the documents (searchability) show a false equivalence between categories of data. A similar point is made by Bingham (2010), for whom the study of newspaper history is significantly impoverished by substitution of plain text for the copy in newspaper archives. The use of keyword searches removes historical texts from the contextual clues needed to decipher them.

However, this is far from a simple technical problem. As Maxwell-Stewart argues:

> (t)he aggregation of data from multiple sources can ... magnify the problems associated with the uninformed mining of archival content in the absence of context. In other words, a failure to understand the political, cultural and social assumptions (and constraints) that shaped the creation and historical use of records will blunt the power of any analytical exercise. Access to ever increasing amounts of digital data will certainly open up research opportunities, but the effective utilisation of that information will rely more than ever on the ability of end users to historicise the data at their disposal. (2016: 361)

The decontextualisation of data also serves to reduce complex phenomena to common data points which serve as their proxies. In everyday life, it is easy to see this trend in the datafication of everything from work performance to health, wherein a given quantifiable measure is made to stand for a more complex category for the purposes of comparison (e.g. step counting for 'exercise'; citations for 'impact'). This is much the concern Livingstone raises in respect of modern-day audience studies. Livingstone argues that the move to BDA has involved a substitution of one rich concept (meaning) by another more measurable but narrower one (behaviour). In the era of BDA, Livingstone argues, the audience is rendered visible only through the by-products of their activity not through sustained scrutiny of their interpretations or the work that they do as audiences. It is as if we are looking to understand an entity by examining only its footprints. The substitution implied here is not only present at the conceptual level. As boyd and Crawford (2012) point out, sub-sets of categories that are more easily envisaged through BDA are made to stand proxy for their broader population. Much research on public opinion, they argue, falls into the error of using 'twitter users' as a proxy for the public in general (2012: 669), for example with all the attendant biases which this implies. BDA, then, in pursuit of a total perspective, complicates the dynamics of the visible, substituting visibility for significance, the track for the tracked.

All of these have implications for the use of archived digital texts, and this is all the more so the greater the remove of materials aggregated, both in time and, I would argue, in 'logic'. However great a gap exists between a 'digital born' news report and a digitised newspaper article, and of course this is a very big gap, both have an orientation on some level to a common logic, of speech to posterity, of the creation of a record, an attempt to inform, to persuade or to emote. Both operationalise a common, if widely divergent, sense of the boundedness of a text, of the types of links and effects it is to allow. These divergencies may be methodologically challenging, but we are at least in the same 'ballpark'. This is not so reliable when broader categories of materials are combined. These aggregations are however the consequence of substituting an empiricist will to totality for a theoretical one. What we gain in coverage, we lose in context. Whereas critical totality de-contextualises social action and re-contextualises it on the scale of the social, BDA de-contextualises the social and relocates it on the level of descriptive, and a particular type of descriptive at that—the associational.

Identity versus Association

The shift towards a research logic focused on association has profound implications, and not least of these is the work of the analyst shifts to a focus on secondary aspects of phenomena rather than phenomena themselves. Relational analyses can examine the interaction between elements but not the elements themselves. We can chart distribution, frequency, adjacency to other phenomena but not the inherent qualities of those phenomena, which becomes unknowable from the point of view of these kinds of analyses.

A good example of this would be the nature of influence. It is not coincidental that the origins of BDA are most often related back to citation analysis (see Merton 1973; deSolla Price 1965; Leydesdorff and Milojevič 2015). These accounts are concerned with the circulation and centrality of texts as a means of determining influence. A good example of this is Spitz and Horvát's (2014) combined analysis of multiple types of citations within film (parody, reference, feature, editing, remaking, follow ups) offering a way to broaden the results of this kind of work, measuring the centrality of canonical and influential texts. They develop a picture of the cultural impact of films which draws together amateur and professional citation and polls, textual citation in extended media and direct citation of works within other films. Other models of influence use diffusion and adoption as key indices, as in the case of semantic network analysis. Large-scale data sets make it possible to track the transmission of common linguistic elements across media texts, or their circulation within a given genre.

However, citation analysis, in taking these as indices, only partly operationalises the concept of 'influence'. We could discuss influence in terms of textual features more inherent to the texts themselves, for example use of rhetoric or of social semiotics. Citation analysis though uses the visual cues of activity as a proxy for the wider phenomenon, with the kinds of consequences we have discussed above. In the case of Spitz and Horvát (2014), for example, combining measures together to produce an aggregate also means collapsing together different understandings of the idea being studied. Influence as it manifests in movie citations by directors is not the same thing as influence measured by recall or popularity, for example.

Moreover, the assumptions of such studies are structural and totalising, namely that there *is* a wider whole which is of greater significance in apprehending the individual text (or person, auteur, actor) than an analysis of that individual would capture. It displaces the subject in favour of a focus on this purported wider whole. Graham et al. (2016) elucidate this point, drawing on Moretti's argument on nineteenth-century literature that close reading of specific texts, even when undertaken on a heroic scale, cannot allow us to apprehend a system as a whole. A 'field this large cannot be understood by stitching together separate bits of knowledge about individual cases, because it isn't a sum of individual cases: it's a collective system, that should be grasped as such, as a whole' (cited Graham et al. 2016: 70). In the same sense, the ways in which artistic, cultural and informational systems are interconnected cannot be established by looking at specific points within them.

This though relies on maintaining two contradictory propositions at the same time: firstly, that there are specific texts (or individuals) within a field which are of great significance (and that therefore it is meaningful to track their citations), and, secondly, that there is a field which exists independently to them. In Moretti's argument, we must assume a wider field of nineteenth-century literature which is 'out there', independent of the activities of authors of the nineteenth century, their critics, readers, the edifice of literary publishing, academic analysis and so forth. Citation analysis then depends on establishing influence by examining the activities of a field as a whole while maintaining there is no inherent identity to the influencer.

Moreover, in addition to emptying the phenomenon of content, the emphasis on association focuses on the link as the unit of significance. However, here again the data lead us to a substitution, of a link for a relationship, and these are not the same thing. Adjacency, proximity or connection will not always translate as co-variance for the purpose of analysis. The fact of something coinciding within one data scrape doesn't translate as those two things being related. However, as Mayer-Schönberger and Cukier (2013) point out, this is not a mere question of implementation failure. It is not the case that a rarefication of the design of a BD study will eliminate this. The logic of BDA itself, they argue, requires us to 'shed ... obsession for causality in exchange for simple correlations: not knowing *why* but simply *what*' (2013: 7). The move from identity to association is then one which substitutes description and coincidence for causality. In so doing, it 'outsources' the burden of explanation to elsewhere within a given research process (see below).

Dislocation/Relocation

Naturally, the advent of BDA has opened up and closed down sets of questions in research. As boyd and Crawford anticipated:

> Big Data creates a radical shift in how we think about research … it is a profound change at the levels of epistemology and ethics. Big Data reframes key questions about the constitution of knowledge, the processes of research, how we should engage with information, and the nature and the categorization of reality. (2012: 665)

In media history, this concern is often parsed into discussions of availability/accessibility. Milligan (2013), for example, observes the ways in which the availability of digital materials draws scholars to an over-emphasis on a small number of publications in historical periodical research, a point also strongly emphasised by Hobbs (2013) in his discussion of *The Times* (UK) historical archive. The danger of the overconcentration on accessible archives lies in the creation and maintenance of a confected significance which overlays historical experience.

The question of significance though is rescaled by BDA in multiple directions, and in particular in respect of which data points (texts, actors, people) are central to a phenomenon. Caplan, for example, in her discussion of work on an early-twentieth-century Yiddish theatre company argues that data-driven methods can 'offer an important corrective to our understanding of what is central and what is peripheral in theatre history' and 'reveal previously invisible patterns about relationships' (2016: 557). The affordance of BDA, for Caplan, is the reduction of the bias found in memoirs towards the subjective accounts of already-significant actors. This is the research 'plus side' of the more negative tendency of BD in the modern era to erode privacy and destabilise the individual's ability to control their public presentations, which takes on a quite different inflection in historical research. Memoirs offer us a post-hoc depiction of a life in which the presenter may or may not recognise key turning points or key influences. BDA potentially offers a corrective to this, allowing us to decentre both the personal accounts of key figures and sometimes those key figures themselves and reconsider the role of others.

Of course, it is equally possible to argue the contrary. As Bingham (2010) notes, the advent of digital methods is potentially a huge boon to biography as it makes it possible to complete a picture of a given personality using lesser-known, lower-profile sources and mentions of the figure. In this way, BDA can round off our understanding, provide confirmation and perhaps challenge our views on a given individual. Interrogating a large data set by using already-familiar landmarks is also attractive, leaving the researcher less 'at sea' than starting with a pattern and trying to find recurrent references and stable 'dry land' within it, but in so doing this may merely reaffirm the centrality of the already canonical. Spitz and Horvát (2014) grapple with exactly this kind of problem in their work on a social network analysis of movie influence (see above).

Moreover, this kind of reconsideration doesn't allow us to simply substitute one text, actor, person for another now 'de-throned' one. Rather it forces us to reconsider what the nature of influence or contribution may be. An analysis of connections between a group of collaborators doesn't give us any insight into the relative contributions of each; being co-present doesn't translate to being involved. This is a variant of the point raised by Maxwell-Stewart in considering BDA in relation to Australian history that not every piece of archival information can be seen as of equal significance (2016: 361). Every connection can be mapped but this does not mean each connection is significant. The artificial significance potentially ascribed to these is an artefact of a more philosophical commitment of BDA, a rejection of random sampling. As Mayer-Schönberger and Cukier (2013) argue, BDA is born of a different logic of information to that of traditional enquiry and one which rejects 'small data' assumptions. For the authors, small data assumptions are those which derive from information scarcity. Sampling, for example, ironically is an artefact of information scarcity insofar as it is a method of interacting with large data sets according to the logic of small ones which emphasises specificity and exactitude. Random sampling though doesn't scale to niches, which is to say that it flattens and generalises. BDA on the other hand 'gives ... an especially clear view of the granular: subcategories ... that samples can't assess' (2013: 13).

What is being pointed to more broadly here is the idea that BDA as a method is, in itself, objective. Data-driven approaches are seen as overcoming the biases and limitations of perspective of the observer, providing an 'Archimedean' position outside of history. However, whilst the data may be objective (though see Criado Perez (2019) on data bias), the analysis is not, and by extension the assemblage of the data is also not, since it is a product of data plus the assumptions and logics which structure it. It may be an empirical totality, but it is one which has the potential to overstate the margins of a phenomenon. It is therefore important to consider the notions and operations of power enshrined in BDA.

Big Data and Power

The question of how big data challenges and reconfigures power and its forms in wider society is beyond the remit of this chapter, but it is important to note that many of the same processes and changes which critics have observed of the impact of datafication outside of the academy are also present within it. Leading thinkers have grappled with the ways in which datafication, digitalisation and a shift to big data logics alter the relationship between academics and their research (Livingstone 2019; Corner 2019; Anderson 2018; Mahrt and Scharkow 2013). Corner has recently reflected on the impact of the incorporation of social media studies into mainstream media studies research and pedagogy. He identifies far-reaching effects on the discipline as a whole:

> return(ing) 'media studies' to at least some parts of that broader agenda indicated in some uses of 'communication studies'. This is an agenda to do with the practical skills and public implications of everyday communication and the orientations and forms, including forms of writing and of the still image, through which information circulates at 'amateur' rather than 'professional' levels. (Corner 2019: 9)

This is true though not only as a result of the participatory logic of social media, which requires us to interrogate professional/amateur discourses, but also in terms of the forms and genres of communication and the ways in which information circulates in social media ecologies. It is also however true of the methods used to examine these phenomena. In the era of big data, researchers have found the value of using crowd—sourcing at multiple junctures within research, both in data gathering/generation and interpretation of these data. However, though such initiatives are potentially democratising, they also have the potential to reinforce boundaries, albeit under the guise of greater equality. As some have noted, there are sharp disparities of power in data-sharing that do not resolve down to a simple question of access. As Andrejevic (2014) notes, 'what individuals can do with their data in isolation differs strikingly from what various data collectors can do with this same data in the broader context of everyone else's data' (2014: 1674).

Additionally, BDA brings about something of a shotgun wedding between media theorists and researchers within the computer sciences. For some commentators, this embeds inequality within the research collaboration. Turkel and MacEachern, for example, warn that 'if you don't program, your research process will always be at the mercy of those who do' (Turkel and MacEachern 2008, cited Graham, Milligan and Weinhart 2016: 59). Processing the kinds of data volume now available does not sit well with the training of historians for whom historical methods centre on the need to extract the greatest possible use from traces and fragmentary data (Graham, Milligan and Weinhart 2016: 1), which then raises a technical inequality of expertise to the level of a broader philosophical one. Finally, although collaboration between 'hard' data science and humanities offers the opportunity to enquire into a new set of research questions, what is less clear is how these questions come to be proposed in the first place. This devolves down into four areas.

Firstly, how the kinds of research questions proposed relate to and are derived from the broader context of the discipline as a whole. Advocates of BDA often seem to envisage the research process in a 'media studies proposes, computer science disposes' way. Data scientists are seen as enablers in the investigation of questions proposed by, and according to the disciplinary needs of, media theorists. The core disciplinary concerns of data scientists are not seen as part of the research design. In this sense, the research is at best multidisciplinary rather than interdisciplinary.

Secondly, there is the celebration of the 'new' in this research. The much-vaunted benefit of BDA is the discovery of previously undisclosed patterns in data (Ewing et al. 2013). This places greater theoretical and analytical weight onto marginal and liminal cases in the discovery of the new. Whilst this fits well with the laudable ambition to decentre the canonical, and the biases implied in that canon, it poorly meets the aim of providing socially purposive research as the analyst is always in search of the counter intuitive. The problem of counter-intuitive findings is well known in sociology, where the idea that if we are explaining the social world, it should be an explanation which is recognisable to the participants in that world is accepted. Media studies, however, retains an attachment to the hidden and obscured, a legacy from its 1970s roots, which can make a virtue of the transcendence of the everyday. In the case of big data, this presents a conundrum, since we have, outside of the formal analytical techniques, no way of ascertaining the central and the marginal, but the method imposes the need to stress the margins for reasons both idealistic and procedural. Moreover, the methods of BDA are somewhat at odds with each other insofar as we aim to discover the counter-intuitive, using this vast data trove to find questions and answers previously un-imagined, but at the same time the activities of interrogating the data require us to already know what we are looking for in order to generate the patterns in the first place (see above).

Of course, this logic of 'discovery' is itself something of an illusion. Humanities are not 'discovering sciences' in the same way that, for example, theoretical physics is. It is possible for physicists to posit a phenomenon 'out there' which the theory and data concur must exist and go and find it (from a lay perspective, the recent imaging of black holes comes to mind), but this is not the case for humanities. There is sadly not an 'out there' to be found in quite the same sense.

Thirdly, we see a shift towards pattern recognition as a source of analytical power. BDA, as noted above, aims at the discovery of previously undisclosed patterns. However, as Steyerl points out, pattern recognition 'resonates with the wider question of political recognition. Who is recognised on a political level and as what? As a subject? A person? A legitimate category of the population? Or perhaps as "dirty data"?' (Steyerl 2016). Steyerl uses the example of a hotel chain which failed to recognise the validity of data patterns as a result of their own ethnic profiling to illustrate the problems of recognition. The problem takes on a different character in historical research though. Historical enquiry often reflects back into the past contemporary concerns, and this is more likely in respect of newly recognised subjectivities and identities, for example eco-activist, feminist and pacificist. These concerns are projected back into the archive, looking for exemplars of this kind of activity or identity to demonstrate a wider salience/continuity. Big data sets are potentially sufficiently diverse as to provide data supportive of almost any kind of claim, but only by substituting a manifestation of the phenomenon for the significance of the phenomenon itself. That traces are discoverable doesn't mean that phenomena recently recognised have had a continuous existence, or that the traces allow us to interpret a modern-day activity as an elaboration or development of this root. As boyd and Crawford point out, this is a common perceptual bias in BDA. 'Too often, Big Data enables the practice of apophenia: seeing patterns where none actually exist, simply because enormous quantities of data can offer connections that radiate in all directions' (boyd and Crawford 2012: 668). Consequently, the results do not speak for themselves but have to be filtered for their salience, a salience which may depend on a subjective assessment of the strength of a relationship discovered. As Barnes (2013) drawing on Sayer argues, '(n)umbers do not speak for themselves but speak only for the assumptions that they embody. Numbers emerge only from particular social institutions, arrangements and organizations mobilised by power, political agendas and vested interests' (2013: 300). The political nature of this process is however obscured.

Finally, there are impacts which derive from the ways in which the practical work of BDA is done, tied to questions around the social and cultural embeddedness of knowledge. One thing upon which all those who have adopted BDA in historical studies are agreed is the excessive, and sometimes startling, work of data preparation involved before analyses start (Caplan 2016; Maxwell-Stewart 2016; Rose et al. 2015). This goes back to the point raised by Brügger and Finnemann (2013) on the difference between the digital native and imported data. Digital 'immigrant' data is not found 'out there', behaving naturalistically. It must be mined by hand from dispersed archives and then reconciled in some way that allows us to see it as equivalent. Meta-data must be devised, classifications added, or existing ones reinterpreted. The process is inevitably messy and prolonged, and the greater investment in data preparation is a powerful incentive to collaborate more broadly, raising in turn questions over the kinds of funding required, and of pressing significance already, the maintenance and distribution of these resources once assembled and the role of the academic in the custodianship of these (Brügger and Finnemann 2013). The effect of datafication on a discipline which, until now, has had a much more formal and institutional separation between the repository and the analyst needs to be examined in more detail than this chapter will allow. It does however mean that there is an institutional 'push' towards using these data sets once this investment is complete. A large resource once established needs to be used, curated and maintained over time.

Moreover, the pressures from within the academy are mirroring those without it. Widespread digitalisation in media history at least is fuelled by its ability to add commercial value. We can see this in a range of examples from over-layering OCR onto primary materials (as in the case of print) or the recovery of artistic works through remastering, which raises questions both legal and cultural around creativity and the creative industries. The 'value added' here is usually depicted as greater access but this needs to be qualified. Digitalisation may broaden the amount of materials to which we have access, considered numerically, but restrict both the range of people able to access it and the kinds of sources accessed. As Maxwell-Stewart

(2016) has pointed out, BDA proposes questions around copyright and ownership which are not easily resolved, making it hard to verify results through replication, for example. The logic of digitalisation is also driven by copyright in other complex ways, for example reapplying copyright to materials whose protection is about to expire. Thus, the adoption of BDA in humanities more broadly is not a consequence of the unfolding of an ahistorical or atechnological logic but is to some extent institutionally and commercially driven.

Big Data, Semantic Network Analysis and Health Communications

So far in this chapter I have considered the nature of big/relational data a priori, in itself. I now want to move on to look at the ways in which these analyses have been used in representations of health and illness, before going on to examine some studies which have used relational and large-scale data approaches to look at anti-vaccination narratives.

Firstly, though, one point to note here is that although smaller-scale content analyses are characteristic of the field, larger-scale and comparative work is not. There is also something of a bias towards the use of print and textual materials in examining these issues. Quantitative content analyses of television drama and fiction are very uncommon, and by far the majority of studies that take up the issue of representations and media in respect of health (based on my reviews of the journals *Vaccine*, *Health Communication*, the *Journal of Health Communication*, *Social Influence*, *Mass Communication and Society* and a substantial portion of those published in *Communication Studies*) refer to media or news only in respect of social media. In part this may be a result of a post- Covid bias, but it is also clear that social media is regarded as a more generally persuasive and pertinent domain of study for health communication and especially immunisation, at least by media scholars (see above for further discussion).

There are relatively few studies using specifically semantic network analysis for the study of vaccine hesitancy, and where these are available they are most usually confined to social media analyses (see, for example, Lyu et al. (2021); Kang et al.(2017); Luo et al. (2021). Featherstone et al. (2020) used the method to look at childhood vaccination and its depiction on Twitter (now X). Ruiz and Barnett (2015) applied a network analysis to looking at HPV information and disinformation in 2015. Stahl et al. (2016) and Rosselli et al. (2016) independently used this approach to model challenges and opportunities in Web 2.0 technologies and vaccine hesitancy. Semantic network analysis has also been used in some health contexts to re-interrogate existing data, for example Kim and Kim's 2015 work on reinterpretation of stem cell survey responses, and to locate and analyse gaps in research where this impacts on provision of health services (Martinez-Garcia et al. 2022). Smith and Parrott (2012) used this method to test the alignment of study subjects differentiated by gender with media narratives in knowledge systems of HPV. Yoo and Lim (2021) found that semantic network analysis was useful in examining news representations of emotions in the Covid-19 pandemic. Yoo et al. (2019) applied the method to a data set generated from the platform Reddit to examine mental health concerns and specifically bipolar and depressive disorders.

The approach has found the most comfortable home in studies of Covid and social media environments where its relative novelty as a method, in addition to the way in which it can be used to focus on emotion and experience as well as opinion, has won it many adherents. Semantic network analysis has particular applicability in the field of health communication where it can be used to track the movement of knowledge systems across different discursive domains. Sentiment analyses, though a basic level of semantic network analysis, offers a way to identify trends in responses to topics and events as well as trends in their occurrence. In so doing, it offers a way of placing different knowledge systems on a level of equality for the purposes of analysis without needing to make an epistemological or theoretical commitment to seeing them so. As a result, it is a method likely to increase in use. However, the kinds of analysis which can be generated from its use are also likely to remain context specific and dispersed. It is hard to imagine a way to adequately use semantic network analysis to compare different phenomena or to unpick underlying and static meaning structures.

This chapter has flagged up some of the ways in which the advent of BDA alters the kinds of questions which it is logical to ask in historical studies. In particular, it has pointed to the ways that big data displaces some questions in favour of others and replaces one logic of engagement with archives and data with another. Of course, a reasonable rejoinder would be to 'live and let live'. Media studies and history are characterised by multiplicity of approaches which do not always complement each other. Why must a warning bell be sounded over big data? As I have argued here,

BDA is a totalising approach which tends to tuck others into its remit. The institutional pressures and commercial forces which surround it create a greater pressure for adoption. Culturally, BDA mirrors, draws force from and in turn reinforces ideas of personhood and identity which are derived from associational models and macroscopic perspectives. The likely consequences of these elements, brought together, is to fundamentally alter the kinds of questions which the discipline proposes.

I will now move on to look at a comparison of the applications of the approaches identified above to the vaccine crisis points of polio and pertussis in the twentieth century.

References

Anderson, C.W. (2018) Apostles of Certainty: Data Journalism and the Politics of Doubt. United Kingdom: Oxford University Press (Oxford Studies in Digital Politics). Available at: https://doi.org/10.1093/oso/9780190492335.001.0001.

Andrejevic, M. (2014) 'The Big Data Divide', International Journal of Communication, 8, pp. 1673–1689.

Aupers, S. (2012) '"Trust no one": Modernization, paranoia and conspiracy culture', European Journal of Communication, 27(1), pp. 22–34. Available at: https://doi.org/10.1177/0267323111433566.

Barnes, T.J. (2013) 'Big data, little history', *Dialogues in Human Geography*, 3(3), pp. 297–302

Bingham, A. (2010) '"The Digitization of Newspaper Archives: Opportunities and Challenges for Historians"', Twentieth Century British History, 21(2), pp. 225–231.

boyd, danah and Crawford, K. (2012) 'CRITICAL QUESTIONS FOR BIG DATA', Information, Communication & Society, 15(5), pp. 662–679. Available at: https://doi.org/10.1080/1369118X.2012.678878.

Brügger, N. and Finnemann, N.O. (2013) 'The Web and Digital Humanities: Theoretical and Methodological Concerns', Journal of Broadcasting & Electronic Media, 57(1), pp. 66–80. Available at: https://doi.org/10.1080/08838151.2012.761699.

Caplan, D. (2016) 'Reassessing Obscurity: The Case for Big Data in Theatre History', Theatre Journal, 68(4), pp. 555–573.

Corner, J. (2019) 'Origins and transformations: histories of communication study', Media, Culture & Society, 41(5), pp. 727–737. Available at: https://doi.org/10.1177/0163443718820666.

Criado Perez, C. (2019) Invisible Women: Data Bias in a World Designed for Me. New York: Abrams Press.

DiCenzo, M. (2015) 'Remediating the Past: Doing "Periodical Studies" in the Digital Era', ESC: English Studies in Canada, 41(1), pp. 19–39.

Erickson, A.T. (2013) 'Historical Research and the Problem of Categories: Reflections on 10,000 Digital Note Cards', in J. Dougherty and K. Nawrotzki (eds) Writing History in the Digital Age. Ann Arbor: University of Michigan Press, pp. 133–145.

Ewing, E.T., Gad, S. and Ramakrishnan, N. (2013) 'Gaining Insights into Epidemics by Mining Historical Newspapers', Computer, 46(6), pp. 68–72.

Featherstone, J.D. et al. (2020) 'Exploring childhood vaccination themes and public opinions on Twitter: A semantic network analysis', Telematics and Informatics, 54, p. 101474. Available at: https://doi.org/10.1016/j.tele.2020.101474.

Graham, S., Milligan, I. and Weingart, S. (2016) Exploring big historical data: the historian's macroscope. London: Imperial College Press.

Hobbs, A. (2013) 'The Deleterious Dominance of The Times in Nineteenth-Century Scholarship', Journal of Victorian Culture, 14(4), pp. 472–497.

Kang, G.J. et al. (2017) 'Semantic network analysis of vaccine sentiment in online social media', Vaccine, 35(29), pp. 3621–3638. Available at: https://doi.org/10.1016/j.vaccine.2017.05.052.

Kim, L. and Kim, N. (2015) 'Connecting opinion, belief and value: semantic network analysis of a UK public survey on embryonic stem cell research', Journal of Science Communication, 14(1).

Leary, P. (2005) 'Googling the Victorians', Journal of Victorian Culture, 10(1), pp. 72–86. Available at: https://doi.org/10.3366/jvc.2005.10.1.72.

Leydesdorff, L. and Milojević, S. (2015) 'Scientometrics', in M. Lynch (ed.) International Encyclopedia of Social and Behavioral Sciences (2nd Edition). 2nd edn. Oxford: Elsevier, pp. 322–327.

Livingstone, S. (2019) 'Audiences in an Age of Datafication: Critical Questions for Media Research', Television & New Media, 20(2), pp. 170–183. Available at: https://doi.org/10.1177/1527476418811118.

Luo, C. et al. (2021) 'Exploring public perceptions of the COVID-19 vaccine online from a cultural perspective: Semantic network analysis of two social media platforms in the United States and China', Telematics and Informatics, 65, p. 101712. Available at: https://doi.org/10.1016/j.tele.2021.101712.

Lyu, J.C., Han, E.L. and Luli, G.K. (2021). 'COVID-19 Vaccine-Related Discussion on Twitter: Topic Modeling and Sentiment Analysis.' Journal of medical Internet research, 23(6).

Madsen-Brooks, L. (2013) '"I Nevertheless Am a Historian": Digital Historical Practice and Malpractice around Black Confederate Soldiers', in J. Dougherty and K. Nawrotzki (eds) Writing History in the Digital Age. University of Michigan Press, pp. 49–63.

Mahrt, M. and Scharkow, M. (2013) 'The Value of Big Data in Digital Media Research', Journal of Broadcasting & Electronic Media, 57(1), pp. 20–33.

Martinez-Garcia, M., Camacho, J. and Hernández-Lemus, E. (2022) 'Connections and Biases in Health Equity and Culture Research: A Semantic Network Analysis', Frontiers in Public Health, 10, p. 834172. Available at: https://doi.org/10.3389/fpubh.2022.834172.

Maxwell-Stewart, H. (2016) 'Big Data and Australian History', Australian Historical Studies, 47(3), pp. 359–364. Available at: https://doi.org/10.1080/1031461X.2016.1208728.

Mayer-Schönberger, V. and Cukier, K. (2013) Big data: A revolution that will transform how we live, work, and think. Boston, MA: Houghton Mifflin Harcourt.

Merton, R.K. (1973) The Sociology of Science: Theoretical and empirical investigations. Chicago/London: University of Chicago Press.

Milligan, I. (2013) 'Illusionary Order: Online Databases, Optical Character Recognition, and Canadian History, 1997-2010', Canadian Historical Review, 94(4), pp. 540–569.

Nicholson, B. (2013) 'The Digital Turn', Media History, 19(1), pp. 59–73.

deSolla Price, D. (1965) 'Networks of Scientific Papers', Science, 149(3683), pp. 510–515.

Rose, S., Tuppen, S. and Drosopoulou, L. (2015) 'Writing a Big Data history of music', Early Music, 43(4), pp. 649–660. Available at: https://doi.org/10.1093/em/cav071.

Rosselli, R., Martini, M. and Bragazzi, N.L. (2016) 'The old and the new: vaccine hesitancy in the era of the Web 2.0. Challenges and opportunities.', J Prev Med Hyg., 57(1), pp. E47–50.

Ruiz, J. and Barnett, G.A. (2015) 'Exploring the presentation of HPV information online: A semantic network analysis of websites.' Vaccine, 33(29), pp. 3354–9.

Sjøvaag, H. and Karlsson, M. (2017) 'Rethinking Research Methods for Digital Journalism Studies', in B. Franklin and S.A. Eldridge II (eds) The Routledge Companion to Digital Journalism Studies. London and New York: Routledge, pp. 87–95.

Smith, R.A. and Parrott, R.L. (2012) 'Mental representations of HPV in Appalachia: Gender, semantic network analysis, and knowledge gaps', Journal of Health Psychology, 17(6), pp. 917–928. Available at: https://doi.org/10.1177/1359105311428534.

Spitz, A. and Horvát, E.-Á. (2014) 'Measuring Long-Term Impact Based on Network Centrality: Unraveling Cinematic Citations', PLOS ONE, 9(10), p. e108857. Available at: https://doi.org/10.1371/journal.pone.0108857.

Stahl, J.P., Cohen, R. and Denis, F. et al (2016) 'The impact of the web and social networks on vaccination. New challenges and opportunities offered to fight against vaccine hesitancy.' Med Mal Infect., 46(3), pp. 117–122.

Steyerl, H. (2016) 'A Sea of Data: Apophenia and Pattern (Mis-)Recognition', E-Flux journal, p. online.

Tanaka, S. (2013) 'Pasts in a Digital Age', in J. Dougherty and K. Nawrotzki (eds) Writing History in the Digital Age. University of Michigan Press, pp. 35–46.

Toscano, A. (2012) 'Seeing it whole: staging totality in social theory and art', The Sociological Review, 60(S1), pp. 64–83.

Wright Kennedy, S., Kuzmin, J.C. and Jones, B. (2017) 'New Methods in the History of Medicine: Streamlining Workflows to Enable Big-Data History Projects', Medical History, 61(3), pp. 477–480. Available at: https://doi.org/10.1017/mdh.2017.54.

Yoo, M., Lee, S. and Ha, T. (2019) 'Semantic network analysis for understanding user experiences of bipolar and depressive disorders on Reddit', Information Processing & Management, 56(4), pp. 1565–1575. Available at: https://doi.org/10.1016/j.ipm.2018.10.001.

Yoo, S. and Lim, G. (2021) 'Analysis of News Agenda Using Text mining and Semantic Network Analysis: Focused on COVID-19 Emotions,"', Journal of Intelligence and Information Systems., 27(1), pp. 47–64.

CHAPTER 4

Polio

Abstract This chapter is a case study of newspaper reportage of polio and polio vaccination programmes in the UK in the period 1955–1960. The chapter uses digitalised archive data and compares the utility of semantic network analysis and qualitative discourse analysis to understanding how the topic came to be framed in the UK press. The chapter argues that vaccination at this period in the UK became tied up with issues around social and political restructuring within the UK regions on the one hand and post-war fears for the future of the UK on a global economic stage on the other. The chapter traces the impact of polio and the fear it engendered on reportage and argues that vaccine hesitancy at this period exhibits a significantly different modality to that of previous and subsequent 'crisis points'.

Keywords Polio • UK newspapers • Semantic network analysis • Anti-vaccination • Metaphors • History

The attempt to eradicate childhood polio through successive vaccination campaigns is, most would agree, the exemplary foundational vaccination 'crisis' point of the twentieth century, a series of events which formed a mould for future immunisation initiatives and bequeathed a legacy of concerns which would reappear again and again over the subsequent decades. It was also, from a British perspective, the point at which vaccination

A. Cavanagh, *Anti-Vaccination and the Media*,
https://doi.org/10.1007/978-3-031-70559-5_4

hesitancy came to be seen as an American phenomenon. As Durbach's (2005) work has detailed, opposition to vaccination in the UK burgeoned in response to the experiences of smallpox immunisation, featuring compulsory immunisation and sanctions, and were filtered through the lens of class conflict, exploitation and the (sometimes) rather heavy-handed reactions to working-class refusal on the part of medical authorities in the nineteenth century (see Introduction for further discussion). This stands at a remove from the concerns evidenced in anti-vaccination discourses in the US, where the fear of direct contamination collided with a populist anti-science agenda, underscored by, and emphasising, regional and racial inequalities in healthcare and medical access and treatment.

However, by the point at which polio shifted from being a background disease of childhood to a virulent public health crisis, anti-vaccination in the UK had become inextricably tied into concerns which were more bourgeois in character. In particular, opposition drew character and focus from one of its key sites of enunciation, the anti-vivisection movement, and also from the political and judicial legacies of the First and Second World Wars. In terms of the latter, as Durbach (2005) further points out, the First World War category of the conscientious objector, with all its moral, cultural and legal ambiguities and inflections, emerged in the first case from legal wrangles around opposition to smallpox immunisation, opening out for contestation the legal and social 'contract' between the citizen and the state where matters of duty were concerned. The category was not reclaimed or reincorporated into anti-vaccination narratives in the post war period as such, but the citizen's right of refusal to meet state demands was, as Hodgson (2022) argues, pathologised and swept up into a broader set of health concerns, largely around post-war mental health (in response to the generation defining mental agonies of those who survived the trenches) and national reconstruction. For a variety of reasons, conscientious objectors, or at least those who successfully maintained their occupancy of the category, tended to be middle class (see Luckhurst 2016, on the representation of the conscientious objector). On the other hand, the anti-vivisection of the movement of the nineteenth century was inherently middle class in nature, formed within the spheres of literary and philosophical societies, social reformers and the medical profession and frequently playing out through journals of note and letters to the editor (see Hamilton 2010). In combination, then, it can be seen that the social bases and audiences for anti-vaccination in the period prior the polio epidemics of the early twentieth century were already very different from those of the nineteenth century in the UK and considerably more at variance than those of the US, to which we now turn.

That polio came to the awareness of the UK public as an 'American epidemic' is at least as much a function of the close relationship between the British and the US press as it was of the origins of the outbreak. Reportage of 'American affairs' was a staple of the UK popular press, as indeed it had been for some time (see Weiner 2012). However, there is no disputing that cases in the US were both startling in their numbers and severity. The period between 1894 and 1916 saw scattered outbreaks of epidemic polio in Europe and the US, but the 1916 epidemic was particularly high profile in the US, especially in New York and surrounding areas. The US had over 27,000 reported cases (41.4 per 100,000) in 1916, as against fewer than 2000 in the previous year (Snowden 2019: 389). It was at this juncture, as Snowden argues, that 'modern polio came of age' (ibid.: 388), which is to say as a virulent disease of mass populations. Polio remained in epidemic outbreaks throughout the 1920s and 1930s, with peaks occurring in 1927, 1931, 1935 and 1940–1941 in the US. From 1943 to 1954, cases remained high, building to over 57,000 (37.2 per 100,000) in 1952.

The starkness of the figures, however, does not account for the dread which the disease inspired in the public imagination nor with the persistent sense of this being an American phenomenon. The latter can in part be explained by the way in which all diseases come to be represented as external threats, with illnesses characteristically represented in nationalised and often racialised terms (see Chap. 2). Moreover, as, for example, Goldstein's study of HIV in Canada suggests, medical distrust lends itself easily to the promulgation of conspiracy theories (Goldstein 2004), and conspiracy theories in turn naturally find rich ground where a sharply defined in/out group is present (see Chap. 5 for further discussion). The 1950s in the UK were particularly fertile ground for a resurgence of anti-American feeling driven in part by anti-populist concerns which equated cultural Americanisation with 'dumbing down' (Jacoby 2008) and acquiring venom from long-standing British resentments about the comparative war experiences of the UK and the US. As Tony Judt has argued, in postwar Europe anti-American feeling stemmed from a relative lack of acquaintance between Europe and the US in all areas save for those of mass media and consumer culture (Judt 2005: 351). It was, then, a sentiment which played itself out almost exclusively within the realm of cultural productions. The relative influence of cultural elites within mass media made their condescension appear more general, and anti-Americanism seem to be far more widespread than it was. For this group:

> America was a land of hysterical puritans, given over to technology, standardization and conformism, bereft of originality of thought. Such cultural insecurities had more to do with the pace of change in Europe itself than with the challenge or threat posed by America. Just as European teenagers identified the future with an America they hardly knew, so their parents blamed America for the loss of a Europe that had never really been, a continent secure in its identity, its authority and its values, and impervious to the sirens of modernity and mass society. (Judt 2005: 353)

The overall cultural environment then was conducive to an unsympathetic portrayal of polio as an American disease in the UK press.

The dread the disease inspired, however, was all its own. In part, this was a result of the demographic afflicted. Unlike in the UK, where the link between poverty, poor sanitation and disease was attenuated by mixed living and working spaces in the older cities (e.g. York and London), wealthy Americans had enjoyed security from many of the diseases of urban poverty that had ravaged nineteenth-century cities (Snowden 2019). Modern polio though was no respecter of neighbourhood boundaries, and indeed appeared to be more prevalent in affluent neighbourhoods. As Millward explains, where communities had access to adequate sewerage systems, a culture of hygiene and the resources to support it, children could avoid exposure to polio in youth but that made them more likely to develop the acute disease later in life (Millward 2019). In this sense, modern polio was a counter-intuitive result of the success of earlier disease reduction initiatives. Polio was also simply more 'visible' in the West, in part because disabled children were, if not an unusual sight, at least a less-prevalent one, in contrast to the plight of children in countries where child health and low mortality rates were less taken for granted (Snowden 2019). The British public also drew its basic models of illness from previous epidemics of tuberculosis and influenza with the result that visible ill health and disability were often stigmatised. Consequently, disabled children could often find themselves largely confined to the household, at least amongst families affluent enough to be able to support them so. The visibility of polio victims, then, shocked the public, a shock which would later be used to good effect in portrayal of polio victims through an archetypal image of a child on crutches with withered limbs, an image that became central to the campaign to research and eradicate polio in the US.

Polio also had a considerable advantage over rivalrous diseases in the sheer capacity to inspire terror. Polio created an overwhelming dependency in those unlucky enough to be severely affected and that dependency was magnified by the key role played by a largely unknown technology, the mechanical respirator or 'iron lung'. The iron lung was first used in 1928 in Boston and is credited with saving thousands of lives of those afflicted by paralytic polio. The machine was in essence a metal box with bellows which worked by adjusting the pressure in the box such that the patient's diaphragm expanded and fell with the operation of the machine. A device both clunky and fragile, it depended for its operation on a stable electricity supply, something which posed a problem in poorer and rural areas where this could by no means be guaranteed. In both the US and Australia, families and communities had to contend with the horror of organising ad hoc volunteer rotas to hand-pump the mechanical ventilators on which severely paralysed victims relied in the event of mechanical failure or an all-too-common outage in the erratic electrical supply (Oshinsky 2006). Initially, the respirators were in short supply, and early reportage of polio in the UK included lurid descriptions of patients requiring prolonged artificial respiration until a respirator could be obtained (see *The Times*, 3 August 1938, p. 12). *The Daily Mail* on 8 December 1957 carried the story of a 'polio mother' who endured an 80-mile 'dash' from Yeovil in Somerset to Exeter in North Devon, whilst a boat battled a gale to bring an 11-year-old from Alderney to Guernsey for polio treatment. In such cases, then, where wealth could confer neither protection nor guarantee the availability of medical technology necessary to sustain life, polio was a disease to inspire terror, especially where the prevalence of the illness was high.

Even when operational and available, the iron lungs were an object of terror, both to their occupants and to their carers (see, Gould 1995; Sass et al. 1996). Tony Gould, a polio survivor describes his experience thus:

> I felt as though both my legs were clamped to the sheet and the iron lung firmly clasping my neck was a hideous form of straitjacket. There were moments when I felt so completely vulnerable that I broke into a cold sweat, when I was so overcome with claustrophobia that my head seemed to burst with the silent screaming of panic. Then I would concentrate on the heavy,

monotonous breathing of the machine which caused my chest to heave and subside. (Le Fanu 1995)

Gould remained in the respirator for a month, but others were less fortunate. *The Times* in 1937 reported on the repatriation to the US of Fred Snite, son of a wealthy businessman, who had been kept alive in an iron lung for 14 months. Snite travelled 'attended by a large staff of doctors, nurses, and electricians. When transferred … to the ship he was removed from the iron "lung" for 3-minutes, during which he was kept alive by a hand-worked respiratory machine. He now speaks freely and can move his fingers. He is surprisingly cheerful' (*The Times*, 7 June 1937, p. 14).

This was surprising indeed given patients' and carers' experiences of the respirator. As Silver and Wilson (2007) point out, after the initial, often-stressful transfer to the machine, patients were obliged to adjust to a frightening level of dependence:

> Patients quickly discovered they could talk only when the machine expelled air from their lungs and conversations became disjointed. Many also discovered that they were completely paralyzed and unable to move a muscle voluntarily. They couldn't scratch an itch, move to a more comfortable position, cover or uncover themselves, feed themselves, or take a drink of water. They were totally dependent on someone else to do all the ordinary things they had done only hours or days earlier. (2007: 38)

In addition, medical staff struggled to provide care. All interactions with the patient were carried out through access holes in the tank in later models, lowering the pressure and the effectiveness of the machine. Nurses therefore had to work swiftly and hospital protocols for containing infectious disease often excluded relatives from visiting. The respirator then isolated the patient as well as emphasising and increasing their dependency. Those who were obliged to use them were often hypoxic before transfer and clinical outcomes, therefore the poorer. The use of the image of the 'iron lung' in fundraising for the National Institute for Infantile Paralysis (collection tins, for example, were designed to look like scaled-down respirators) reproduced this fear outside of immediate medical settings. Patrick Cockburn describes the atmosphere of terror around polio and the 'iron coffin' in Co. Cork in Ireland in the 1950s, a fear which in this case was compounded by the reputation of and authoritarian regimes at the 'Fever' hospitals in Cork.

In part we can see this terror, both of polio and later of its vaccines, as tied to a fear of the technologically strange, and of the breach of boundaries implied by both the vaccines and the treatments necessary to sustain life. The theorist Donna Haraway, in her landmark work 'cyborg identities' (1991), points out that fear of technology is produced where there is a breach in the cultural boundaries between, on the one hand, human and animal, and on the other 'animal-human' and machine. Polio breached both, on the one hand creating a human–machine subject through the dependency of the polio victim on the ventilator and, on the other, evoking the horror of the human–animal. In so doing, polio concretised and revivified long-established fears of bodily pollution that had been stoked by the UK anti-vaccination movements origins in anti-vivisection. Thus, for example, *The Telegraph*'s coverage of the UK's vaccine campaign emphasised the animal origins of the vaccine:

> Macleod acknowledged that 'Many interpretations have been put on the attitude of this country to this development. Perhaps it is as well to out the facts in plain words. This new vaccine involves inoculating our children at repeated intervals with a preparation derived from the kidneys of dead monkeys. We are carrying out intensive tests as to the exact effects so that we can eliminate any possible dangers from it'. ('Polio Vaccine Supplies in Autumn', *The Daily Telegraph*, 26 April 1955)

Taken as a whole, *The Daily Telegraph*'s coverage erred on the side of suspicion of the Salk vaccine when it became available. The reportage of the 1955 World Health Organisation's conference, where the recommendation to bring forward the vaccination programme was made, for example, was carried under the headline *Caution on Polio Vaccine Urged* (28 November 1955). The pioneering research which allowed monkey kidney to be substituted for human placental tissue 'salvaged'[*sic*] from any maternity hospital ('Growing Polio Virus: Method without Monkeys' *The Daily Telegraph*, July 1955) was acknowledged only briefly, clearly not evoking the same concerns. The fear that vaccines breach the bodily autonomy of the recipients is of course a commonplace of all anti-vaccine narratives. As Coffelt and Djandji (2023) argue in respect of Covid-19, viruses are themselves often represented through the metaphor of the fictional Frankenstein's monster, a process which simultaneously 'others' the virus and its carriers/sufferers and points to its 'uncanny' or 'mutant' nature.

The terror of polio would not start to subside, in popular culture at least, until the end of the 1950s, and it was at the point that polio once again became a disease of the poor (see Conis 2015). Successful vaccination campaigns with the Salk vaccine disrupted the cycle of epidemics, leaving pockets of the disease circulating largely in communities either unreached by, or opposed to, vaccination. The development of the Sabin vaccine was crucial in reducing these 'pockets'. The novel mode of 'sugar cube' administration had multiple benefits in this. The dose could be given cheaply, the training of vaccinators was cost-efficient and brief, and since no injection was needed, there was less public resistance. An unanticipated though equally significant benefit of the method was the rhetorical distance it offered from the Salk vaccine in the years after the tragic 'Cutter incident' (1955), where a manufacturing error led to the administration of live vaccine in error in the US, resulting in 40,000 direct infections with 51 cases of paralysis and 5 fatalities, and generating a short-lived epidemic outbreak which resulted in further cases of paralysis and death (see Oshinsky 2006). Sabin's sugar cube did not carry the same desperate connotations in the popular imagination (Oshinsky 2006).

In the UK, the story of polio and its vaccines was very different and focused in on a more amorphous yet more pressing set of national interests, and it is to these that we now turn. Initially, this discussion uses semantic network analysis in order to bring out the stories of the coverage, the implicit and underlying assumptions which came to stand proxy for the 'common sense' of the vaccine. The corpus of materials which were generated from the newspaper archives were analysed using this method (see Chap. 2). This allowed a fruitful comparison of how particular terminology changed in inflection over time and between publications which can be used to sketch the transformations in representation between periods. As noted earlier, the aim was not to understand how the coverage of specific viruses and their vaccines developed over time using this method, for that the qualitative discursive materials were required, but rather to capture a snapshot of the themes attendant on this virus 'crisis point' in order to compare with others (see Chap. 5 on pertussis). The data were separated by newspaper to affect a comparison and as a means of corroborating the analysis.

In the case of the polio 'crisis point', the two graphs (*The Daily Telegraph*, Fig. 4.1, *The Daily Mail*, Fig. 4.2) which were derived share central thematic concerns which merit further analysis using the qualitative narrative and metaphorical analysis, as discussed in Chap. 2. The first and most significant of these was a combination of nationalism and intense distrust of American science. From the outset, polio in the British press was a story 'about' nationalism, patriotism and the UK's position on a world stage in the last days of empire on the one hand, and of class and the failing cultural authority of a mandarin class whose visions were at such odds with the bright horizons opened up by new technology on the other. As Millward argues, the 'British political classes were already anxious about Britain's status as a fading power, with science and technology a potent symbol of this decline in the nuclear age…The pharmaceutical industry was one area where Britain might be able to compete' (Millward 2019: 124). In the race for a vaccine, and more significantly the race to

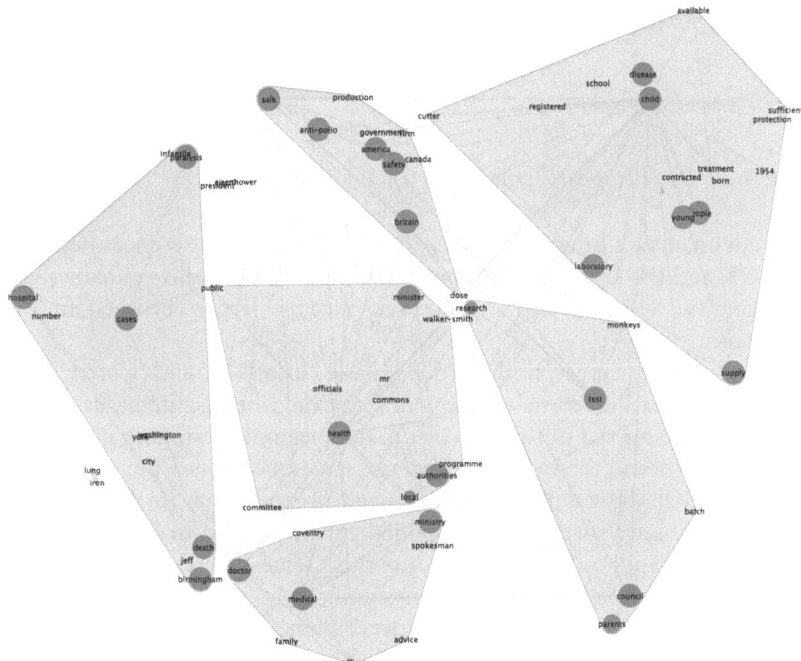

Fig. 4.1   *Daily Mail*/polio

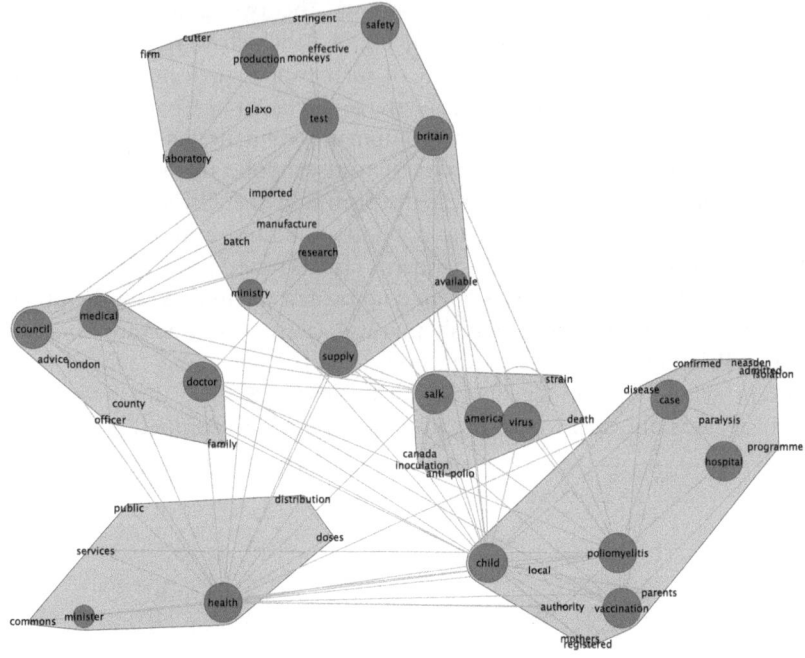

Fig. 4.2   *Daily Telegraph*/polio

provide manufacturing capacity, coverage focused on differences between the UK and the US, with European reactions to the polio epidemic, not to mention a European-manufactured vaccine, largely eclipsed in the coverage.

As we would expect, both graphs feature clusters around 'Cutter', and 'laboratories' as linked to 'Cutter', and it would be disingenuous not to recognise the impact of this incident in undermining/bringing into question confidence in the vaccine. However, it is not evident from the British coverage that this confidence was indeed present prior to this. On the contrary, we see a marked and sustained anti-Americanism as a key theme in coverage of polio. However, the way this theme appears in the coverage by the different papers is subtly but significantly inflected. In *The Daily Mail*'s 'America' cluster, 'safety', 'government', 'firm' and 'production' are the key terms, along with 'Britain'. This evidences concerns, shared between the papers in this sample but differently inflected, about the quality of the imported vaccine. In the case of *The Daily Mail*, this was more tied to nationalism and anti-Americanism than that of *The Daily Telegraph* (see below). The intermediary terms ('firm', 'production') highlight the emerging sense that vaccine development in the US is driven by a for profit motive.

In this we also see the early emergence of 'big pharma' as the logical villain of the piece, a narrative which, continued through into the era of Covid-19, has been so foundational to anti-vaccination movements. The adjacency of 'government' and 'safety' to 'America' in *The Daily Mail* corpus is also significant, pointing to the comparison effected by the paper between American 'commerce' and the UK's formal political response to the same. One of the features of semantic network analysis which make it of particular use in this kind of work is being able to look not only at the co-occurrence of terms but also at the way some terms are 'appropriated' into other clusters. In the case of *The Daily Mail* here, we can see that whereas 'government' appears in the 'America' cluster, logically corollary terms appear elsewhere in the graph. Thus, 'Washington' appears in the 'cases' cluster, whereas 'officials' and 'commons' (the UK House of Commons here) appear most prominently in the 'Health' cluster.

In the case of *The Daily Telegraph*, by contrast, we can see a clustering of terms around 'America' with 'Salk', 'strain' and 'death'. The 'America' cluster focuses more on the virus itself, with 'industry' terms, for example 'imported', 'manufacture', 'production', being attached to the 'test'/'research' cluster. The concern over the imported vaccine for the *Daily Telegraph* is strongly tied to 'Salk', pointing to a greater differentiation in the variety of vaccine than in the case of the *Daily Mail*. The fact that the term 'virus' itself appears as part of the 'America' cluster is also revealing. *The Telegraph* does portray polio itself as an American 'problem' in the first instance, but we can also note that the 'contagion narrative' described by Wald (2008) is also evident here, with 'new' polio being presented as an American virus (see Chap. 2, this volume).

The second key theme here is localism and autonomy. In contrast to the US, where supplies of vaccine were organised through state infrastructures by an ad hoc group of organisations, medical and health initiatives, in the UK the provision and co-ordination of the 'rollout' was directed centrally by the Ministry of Health and cascaded down to local health authorities and officials. Fears about the quality of imported vaccine and issues with the UK's manufacturing capacity came together to produce a scarcity of vaccine supplies. The co-ordination of immunisation therefore involved allocation of supplies based on deemed priority. For the UK press, this meant that there were two key areas of concern. In the first instance, the way that the supply was prioritised risked exacerbating local and regional inequalities and long-standing grievances, an issue to which we will return in Chap. 6 on Covid-19. In the second instance, the prioritisation of recipients significantly undermined the local autonomy usually enjoyed by health councils and local medical officers. It is significant in both the qualitative analysis and the semantic network analysis that parental authority/autonomy rarely appears in these discussions other than as an adjunct to concerns around the erosion of local government authority.

The way these themes were elaborated is seen most clearly in the case of the coverage from *The Telegraph*, where two distinct clusters form around terms associated on the one hand with 'poliomyelitis' and 'vaccination' and on the other with 'medical' and doctor'. In the case of the former, adjacent terms include 'local', 'hospital', 'parents', 'child', 'authority' and 'cases', whereas, in the case of the latter, significant terms include 'advice', 'county' and 'officer'. 'Medicine' and its adjacent terms are associated with national and professional organisations, whereas 'vaccination' appears most often and has a stronger association with locally facing institutions of medical governance and delivery. The final cluster around 'health' also elaborates upon this understanding with adjacent terms 'minister', 'services', 'public' and 'distribution'. In 1950s Britain, 'health' was most closely tied with questions of policy and provision, rather than, as it would later become, an artefact of the individual or of the community. It is notable that this echoes and elaborates on the way in which medical doctors were encouraged to think about their professional roles at this period, in a way which observed a distance from patients, distrust of the new National Health Service (NHS) and an orientation to the support of established sources of medical authority. In part, this brittleness was, as a professional tactic, an outcome of the contestations which attended the birth of the NHS, wherein medical doctors' opposition to becoming 'state workers' was eventually overcome only at the cost of permanently souring relations between health authorities, government and the General Medical Council (see Saunders 2022).

In the case of *The Daily Mail*, there is something of a different focus with 'medical' and 'doctor' appearing closer to 'ministry' and 'spokesman'. This is likely to be an artefact of the paper's greater reliance on official spokespersons and government sources as against *The Daily Telegraph*'s more diverse professional sources (see also Dew 1999). This however is a matter of nuance, with both sets of data evidencing the gap between terms referencing medical personnel and those referring to the community experience of either the virus or the vaccine provision. The division between the local as a theatre of delivery and the national as the sphere of advice was part of the tacit common sense of health communication at this time. This is not unexpected. As Wald has noted in a different context, the narratives which attend outbreaks of contagion of all forms act to specify and delineate different communities, creating a greater sense of affiliation and commonalty on the one hand whilst on the other reinforcing and recreating community differences. The experience of an epidemic can 'evoke a profound sense of social inter-connection: communicability configuring community' (Wald 2008: 12). She continues, the 'interactions that make us sick also constitute us as a community. Disease emergence dramatizes the dilemma that inspires the most basic of human narratives: the necessity and danger of human contact' (ibid.: 2). A key element of this community formation is the recognition and isolation of the 'stranger' or the 'outsider', the identification of whom stems from the initial 'anchoring' in Moscovici's terms (see below and Chap. 2). In the case of polio in the UK, these community boundaries were drawn less along the lines of the vectors of the disease itself and more with reference to the agencies involved in the attempt to eradicate it. Antagonism and concern developed along existing social lines of tension.

Thus, the two key themes emerging from this corpus of materials are inflected by the horizons of the bureaucratic and procedural on the one hand and national ambition on the other. These will be discussed in greater detail below. In order to capture the more nuanced aspects of the representation, as noted in Chap. 2, I use a combination of techniques derived from social representations theory on the one hand and critical discourse analysis on the other. This enables a focus on the meanings generated here by foregrounding their development in and between opposing groups. As noted in Chap. 2, the use of critical discourse analysis in historical contexts is stronger in analysing modalities of discourse as against analyses of metaphor or other formal aspects of language given the problem of substitution and the attendant difficulties of fixing sets of adjacent or substitutable terms.

THEME 1: ANTI-AMERICANISATION AND PATRIOTISM

Predictably, enough coverage of US produced and sourced vaccines tended to adopt a rather lofty position. Early pieces from *The Telegraph* and *The Daily Mail* focused on the need for any US vaccines to undergo further UK testing before administration could be considered. *The Telegraph* announced: 'Britain to make tests of U.S. Polio Vaccine: Doctors' Cautious Attitude To "Dead Virus" Formula' (21 April 1955) before detailing the plans in motion to allow production of the Salk vaccine to come under the auspices of UK manufacturers, backed by government support and oversight. The subsequent reportage focused on the testing regimens to be employed. Under the headline 'Britain Makes Polio Vaccine' (26 April 1955), the paper outlined hopes that 'all tests would be finished by the autumn and quantities would be available for inoculation by the end of the year. The Medical Research Council is soon to carry out trials in various cities and towns ... Among the "guinea pigs" will be students at Manchester University in addition to children'. The paper also carried coverage of the somewhat equivocal reception of the Salk vaccine by Iain Macleod, minister for health, in the Commons. Macleod, whilst congratulating Salk on what he described as a 'momentous and historic advance', was at pains to reassure the House that administration was by no means assured, even given the agreement in place to buy the output of the two UK manufacturers involved (at this time Glaxo Laboratories and Burroughs Wellcome and Co.). Rather, the Ministry for Health was intending to wait for the seasonal cases to fall back and further testing to establish whether a vaccine intended for the US would be effective in the UK ('Britain to Make Tests of U.S. Polio Vaccine', *The Daily Mail*, 21 April 1955). *The Daily Mail* also was at pains to evoke the need for loyalty, even in the face of fears around the quality of the vaccine:

> We have all been frightened by the Salk vaccine in America. Whatever the new formula may or may not do. we at least owe to our health authorities the refusal to be stampeded by exaggerated claims from the U.S. They have applied to their own stuff the most exhaustive tests. Of course, there is an element of risk, as there must always be in a new treatment—or any treatment. But can anyone believe that any Government. backed by the finest medical advice in the country, would give their sanction unless they were convinced that it was good? (*The Daily Mail*, Friday, 9 March 1956)

These concerns did however provide impetus for the production of the UK's own vaccine, overseen by the Joint Committee on Poliomyelitis Vaccine, to cater for demand both in the UK and the Commonwealth, a challenge to the US' potential monopoly in this area. However, this itself had something of a knock-on effect on the availability of the vaccine. The joint focus on providing for the UK and establishing a commercial stronghold in healthcare in the Commonwealth aggravated the supply issues by which the programme was dogged (Millward 2019) and exacerbated the problems of co-ordinating supply and demand in the distribution of the vaccine. The higher standards of testing required in the UK as compared with the US and Europe also raised the cost of the vaccine outside the Commonwealth, making it uncompetitive internationally.

By 1956, the supply gap in the UK had sweetened UK government's views of the Salk vaccine. Offers to establish supply links with the Pasteur Institute in France raised 'serious doubts on technical and scientific grounds', and caused a minor diplomatic scuffle which was lovingly documented by the papers. The Salk vaccine moved up to be seen as a means to 'fill the gap' until the British supply could be brought up, but *The Daily Telegraph* hastened to reassure the readership that it would 'be subjected to the same stringent tests of safety, potency and purity as British vaccine. Parents who object to its use', the piece continued, 'can have their children vaccinated with British vaccine later' ('Britain to Buy Salk Vaccine', *The Daily Telegraph*, 12 September 1957).

Although editorial pieces continued to strike a tone of scientific remove, around the edges a greater diversity of views were starting to come through, and in large measure these were associated with the then so-called women's pages. An excellent example of this is a profile piece on Margaret Agerholm, polio researcher, which appeared in *The Daily Telegraph*, under the pseudonym Felicia Lamb (likely in reality writer and socialite Lady Pansy Lamb), whose copy generally focused on matters of domestic life, fashion and occasional profiles of working women. The piece traces the work of Agerholm and describes the frustration experienced by those who advocated for speedier provision with the Salk vaccine. Lamb describes Agerholm as having

> spent the last six months trying to find out why there has been a hold up on Salk to implement the inadequate supplies of our own variant. She says that if the story of the fight to get protection in England wasn't tragic, it would be funny. She has battered on the doors of the Ministry of Health and the

Medical Research Council and deliberately stirred up the Press. She took me round the wards of the Nuffield Orthopedic Centre in Oxford where she works, telling me how polio had not only paralysed each patient but also destroyed a family's life. "I want you to know how terrible it is." she said. "Too many people have been saying that the polio rate is comparatively low in this country and we mustn't get emotional about it. If the Americans hadn't got emotional about it, things wouldn't have happened so quickly there".

In a box summary to the piece, Lamb sympathises with parents 'muddled by conflicting accounts'. 'Obviously', she soothes, 'it would have been nice for national prestige if we could have produced all the vaccine we needed, but we couldn't. The quicker we get extra supplies the better' (Lamb 1957).

The ruffling of British pride then seems to have gone a long way towards the amplification of the risk of tainted vaccine and the perpetuation of fear long after the Cutter incident itself. It is certainly true that figures on the rates of infection in countries where the US vaccine was adopted did not exhibit the same spikes as in the UK. However, it is also the case that the virulence of the virus was of less significance, in the UK press at least, than the autonomy of a local response to it, to which we now turn.

THEME 2: LOCALISM AND AUTONOMY

One of the features of the distribution of the polio vaccine in the UK was the local and piecemeal nature of the response. Divided into quasi-autonomous health boards, healthcare and medical provision differed at this time widely across the regions. These differences had a long tail legacy which would come back to haunt health policy in subsequent vaccine programmes insofar as they served to exacerbate and compound narratives of scarcity, and framed access in competitive terms. What this meant in the case of polio though was rather different. The primary way in which 'autonomy' was understood in respect of polio was the local area's right to decision-making, as opposed to a focus on parental or individual rights. Individual areas of scarcity were presented in the press in terms unique to them; seldom were they seen as an exemplar of a wider pattern. Thus, for example, when concerns were raised around the day-to-day running of Neasdon Hospital in Middlesex, a centre for the treatment and quarantine of infection, and Red Cross volunteers were called to assist, the media coverage focused on defusing the sense of alarm, giving prominence to official sources (in this case, A.C.T. Perkins, medical officer for health for Middlesex) and stressing that new hires recruited to handle the larger numbers of patients were only temporary. Perkins was also given space in *The Telegraph* to refute a statement attributed to his department that it regarded the present polio epidemic as the "most serious ever experienced in the country". He said he was unable to trace the source of the statement which was achieving some traction in the UK press. The paper did pause however to point out that '(o)wing to the severe pressure of work caused by the polio outbreak, two women doctors at Neasden have been ordered to rest' ('*Polio Aid by Volunteers*', *The Daily Telegraph*, 20 September 1955).

Very few articles in either paper approached questions of autonomy and choice with a parent's perspective in mind. Where parents and their knowledge and consent were discussed, this was usually as part of a related but peripheral set of concerns. As detailed above, one example of the appearance of these was the women's pages where the aim was, as in the piece by 'Felicia Lamb', to dispel parental confusion, addressing what we would now see as an information deficit, an aim it is not surprising to find is here restricted to reportage specifically aimed at women/mothers.

Parental consent was also presented as a concern of those who had a broader agenda within the health service and wider community. A good example here is the coverage in 1956 of presentations to a British Medical Association conference which brought together various impressionistic and largely speculative concerns and options around vaccination. Under the headline *Polio Risk in Taking Aspirin, Says Doctor* (*The Daily Telegraph*, 11 July 11 1956), the correspondent detailed one doctor's concerns that use of aspirin could make patients more likely to contract illnesses including polio, as it could mimic the effects of steroids given to monkeys in laboratory testing of vaccines. The paper also gave space to the same speaker's recommendation for the introduction of 'German Measles tea parties' and camps, to be organised through schools and the Ministry of Health, before concluding with a discussion of the plight of so-called Pincushion Babies, whose parents, one former medical officer of health had claimed, 'could be forgiven for thinking that the family or public health doctor regarded a child mainly as a vehicle for sticking needles into. ... Parents are going to strike one of these days against the number of times they are asked to bring children to the public health clinics to have needles stuck into them'. Although the piece is straight reportage, its overall tone is sceptical, and the unresearched views of the doctor advocating for spreader events and restricted use of aspirin and those of the former medical officer on the exhaustion of parental consent are placed in the same category. *The Daily Telegraph's* correspondent stops short of mocking the views put forward but distances themselves, giving the final word to the same former medical officer: 'One of the great advantages of being a medical MP is that you can talk as much bilge as you like and no one knows you are talking nonsense'.

The frustrations of low supply of the UK-produced vaccine were managed by staged deliveries to different age groups, relying on parents to register children as they became eligible. However, ultimately this even further exacerbated supply issues as parents responded inconsistently and often only in response to local outbreaks, which, in any case, the UK vaccine was unable to ameliorate (Millward 2019). In addition, local authorities were sometimes unwilling to concede power over local health to central government. This reached a pitch in the case of Burton-on-Trent, where local officials refused to gather the data and permissions necessary for children to receive the vaccination, citing the Ministry of Health's 'cock-eyed methods of organising the vaccination':

> "We were asked to collect the names of children whose parents agreed—but we were told at the same time that only enough vaccine to treat one child in five was available". "We were told that the selection of children would be made entirely by the Ministry. And finally we were given no guarantee that the vaccination would work." It seemed logical to us that the best place to use the vaccine would be in the area where the epidemic broke out But -under the Ministry scheme this would have been impossible. ('A Town Bans Polio Drug for Children', *The Daily Mail*, 8 March 1956)

The town's refusal to comply with the scheme was excoriated in the press for that singularly English offence of presumption. For *The Daily Mail* the Health Committee of Burton-on-Trent was to be seen as arrogant, impertinent and 'know-alls' whose directives were 'egalitarianism run mad' ('Town Hall Know-All', *The Daily Mail*, 9 March 1956). In particular, the *Mail* singled out the council's cited concern over vaccine effectiveness. 'Most people would sooner rely upon the Minister of Health in such a matter than upon the Health Committee of Burton-on-Trent. It is for parents to decide whether their children shall or shall not be vaccinated. By what right do Alderman Clark and his colleagues deny ...choice' (ibid.). The modalities here are complex. Against the medical board (and Alderman Clark in particular), the *Mail* arrays the government, medical advice and parents. The medical board are ravelled into the category of those who may be 'stampeded' by exaggerated claims, and specifically Americans who may be so, bringing the additional force of British elitism to bear.

Neither was *The Daily Mail* alone in these concerns. In the reportage of Burton-on-Trent's stance, as later with Wakefield in West Yorkshire, the concerns highlighted by *The Daily Telegraph* were over the relationship between local and national government, with local councillors' apparently 'cavalier' appropriation of domains of the state accentuated by an appeal to parental rights. 'It was for the parents of Burton to decide and not the representatives of the health committee' ('Town Accepts Polio Vaccine', *The Daily Telegraph*, 12 December 1956). In this body of reportage, parental rights and concerns appear largely as rhetorical flourishes to substantiate the inappropriateness of the regional councils' decisions. It is fair to say that this is quite representative of discourse around polio vaccination in the UK generally. It was far more tied to administrative and supply issues than the fervid technological fantasies to which the political climate of the mid-twentieth century in the UK was conducive, fantasies which culminated ultimately in the 'white heat' rhetoric used to such good effect to recapture Labour politics by Harold Wilson in 1963. Although, as noted above, fears of technology were frequently figured, the sense of technology as a source of solutions was largely absent in the case of the vaccine.

The coverage of the vaccine campaign in the UK then was centrally tied to questions of priority and control. These two themes are elaborated repeatedly within the coverage, a set of concerns which can be understood in the context of the changes which this period sees in the establishment of the National Health Service (NHS) and the changing bureaucratic and professional boundaries, identities and areas of expertise that this entailed. The vaccine programme breached long-entrenched and fiercely defended areas of authority and influence, insofar as it obliged local health services to defer to a national set of regulations and procedures at the same time that the discursive climate around the vaccine itself and the worries about its safety were so pronounced. The fear that the health councils possessed only 'responsibility without power' in the face of the extraordinary flowering of a previously known disease into an epidemic underpins a lot of these concerns. These are reinforced by the wider discourse of anti-Americanism and worry about the cultural impact of American financial and trade dominance in the post-war period. That the UK could step forward on a world stage to regain some of her lost supremacy was an enticing vision, but the more conservative safety testing standards in UK pharmaceutical production put this at a remove. There was no way to compete within the pharmaceutical market without reducing the UK's rigorous testing standards and the considerable frustration this engendered can clearly be seen within the coverage. The notion of a less-well-tested vaccine taking precedence largely as a consequence of the UK's compromised and limited production capacity was a bitter pill to swallow, and this accounts for much of the anti-US rhetoric. The UK's somewhat cavalier dismissal of the offer of French vaccine stocks carried none of the sense of emotional investment as that of the US.

In the case of polio, the way that the vaccine was presented not only followed the lines of an archetypal Britishness but served to reinforce these. Modern polio was a 'democratic' if cruel disease, impassively afflicting any sufferer. Neither were the resources to ameliorate its effects subject to vectors of distribution with which the UK was familiar. The vaccine programme however was built on, and reiterated, long-entrenched community inequalities and social concerns.

References

Coffelt, A. and Djandji, A. (2023) 'Mutant metaphors: Frankenstein in the era of COVID-19.' Med Humanit, 49(2), pp. 272–277.

Conis, E. (2015) VACCINE NATION: AMERICA'S CHANGING RELATIONSHIP WITH IMMUNIZATION. Chicago/London: University of Chicago Press.

Dew, K. (1999) 'Epidemics, Panic and Power: Representations of Measles and Measles Vaccines', Health, 3(4), pp. 379–398. Available at: https://doi.org/10.1177/136345939900300403.

Durbach, N. (2005) Bodily Matters: The Anti-Vaccination Movement in England, 1853–1907. Durham and London: Duke University Press.

Goldstein, D. (2004) Once Upon a Virus: AIDS Legends and Vernacular Risk Perception. Logan, Utah: Utah State University Press. Logan, Utah, US: Utah State University Press.

Gould, T. (1995) A Summer Plaque: Polio & its Survivors: New Haven and London: Yale University Press.

Hamilton, S. (2010) 'Reading and the Popular Critique of Science in the Victorian Anti-Vivisection Press: Frances Power Cobbe's Writing for the Victoria Street Society', Victorian Review, 36(2), pp. 66–79.

Haraway, D. J. (1991) Simians, cyborgs, and women : the reinvention of nature. New York: Routledge.

Hodgson (2022) 'Pathologising "Refusal": Prison, Health and Conscientious Objectors during the First World War.', Social history of medicine : the journal of the Society for the Social History of Medicine., 35(3), pp. 972–995.

Jacoby, S. (2008) The age of American unreason : dumbing down and the future of democracy. London: Old Street.

Judt, T. (2005) Postwar: A History of Europe since 1945. New York: Penguin.

Lamb, F. (1957) 'The People who Fight Polio', The Daily Telegraph, 26 September, p. 9.

Le Fanu, J. (1995) 'Victory over the plague of youth', The Times, 6 April, p. 13.

Luckhurst, T. (2016) '"The Vapourings of Empty Young Men?"', Journalism Studies, 17(4), pp. 475–489. Available at: https://doi.org/10.108 0/1461670X.2015.1071196.

Millward, G (2019) Vaccinating Britain: Mass vaccination and the public since the Second World War Manchester: Manchester University Press.

Oshinsky, D.M. (2006) Polio: An American Story New York: Oxford University Press

Sass, E., Gottfried, G. and Sorem, A. (1996) Polio's Legacy. Lanham, Maryland: University Press of America.

Saunders, J. (2022) 'The making of "NHS staff" as a worker identity, 1948-85", in J. Crane and J. Hand (eds) Posters, Protests and Prescriptions: Cultural Histories of the National Health Service. Manchester, UK: Manchester University Press.

Silver, J. and Wilson, D. (2007) Polio Voices: An Oral History from the American Polio Epidemics and Worldwide Eradication Efforts. Connecticut: Praeger (The Praeger Series on Contemporary Health and Living).

Snowden, F.M. (2019) Epidemics and Society. Yale University Press. Available at: https://doi.org/10.2307/j.ctvqc6gg5.

Wald, P (2008) Contagious Cultures, Carriers, and the Outbreak Narrative Durham, NC US: Duke University Press.

Weiner, J. (2012) The Americanization of the British Press, 1830s-1914: Speed in the Age of Transatlantic Journalism. Basingstoke: Palgrave Macmillan.

CHAPTER 5

Pertussis

Abstract This chapter is a case study of newspaper reportage of pertussis/ whooping cough vaccination programmes in the UK in the period 1974–1979. The chapter uses digitalised archive data and compares the utility of semantic network analysis and qualitative discourse analysis to understanding how the topic came to be framed in the UK press. The work identifies the growing influence of formal campaigning and in particular the creation of communities of common interests around illness/ disability and vaccination hesitancy. The chapter traces the early history of public relations techniques in anti-vaccination campaigning. The chapter argues that the case of whooping cough laid the groundwork for a codification of the responsibilities of citizens to the state in the matter of health.

Keywords Pertussis • Whooping cough • UK newspapers • Semantic network analysis • Anti-vaccination • Metaphors

The pertussis vaccine controversy was, as Baker has argued, 'the most significant setback for the cause of immunisation since the smallpox vaccine debates' (2003: 4003). This so-called 'vaccine revolt' (Baker 2003) in the 1970s and the early 1980s in the UK was provoked by a collapse in confidence in the vaccine programme, which led to a series of whooping cough epidemics in the UK, threatened the effectiveness of the successful polio

© The Author(s), under exclusive license to Springer Nature Switzerland AG 2024
A. Cavanagh, *Anti-Vaccination and the Media*,
https://doi.org/10.1007/978-3-031-70559-5_5

vaccination programme also in the UK, and was briefly 'exported' as a concern to the US, where it led to a rise in litigation which in turn threatened to undermine the 'vaccine narrative' (Baker 2003: 4007). It stands as 'a "lesson from history"... (for)... public health professionals. It is seen as a good example of how a mature vaccination programme in a high-income setting can undergo a loss in public confidence' (Millward 2019: 169–170). The pertussis controversy marked a point at which the validity and safety of vaccines sui generis was brought into question. Many countries experienced a loss of public or scientific confidence in particular vaccines without that impacting directly on the uptake of vaccines more generally. This then is the point at which modern anti-vaccination as a coherent and systematic set of beliefs begins to emerge.

As Millward further argues, the pertussis crisis was crucially tied up with at least two sets of modern concerns. The first was the extent of the boundaries and obligations of the state to its citizens. The question of how far individual citizens should be expected to place themselves at hazard for the benefit of a wider community was central to the discussions. Moreover, the pertussis vaccine revolt was, he argues, about the place of risk, considered here as the ways in which modern societies assess, identify and manage risk (see Beck 1992), in the work of medical authorities. The pertussis revolt brought two different modes and orientations in managing risk into collision. On the one hand, it is a given that the state's duty to avail the community of the best immunological advances available (given inequality of access, wealth disparities and so forth) can come into conflict with the equal duty to prevent harm to individuals. This would be true of any public health measure; however, in the case of pertussis the specific nature of the understood threat from the vaccine presented this knotty problem in an unusual way. Pertussis itself was a well-known illness with a low rate of complications and fatalities amongst older children. Family doctors, as Baker notes, were often imprecise in diagnosing it, labelling any prolonged respiratory illness in childhood as pertussis. Colloquially, the illness was known as whooping cough and known, in the UK and elsewhere as a 'normal' childhood ailment (see Gangarosa et al. 1998: 359). Thus, unlike polio, which had appeared in dramatic epidemics and presented as a novel threat, pertussis was familiar, known, part and parcel of childrearing. That is not of course to say that pertussis infection was benign in itself, but rather that it was a familiar worry. The threat in this case comes directly from immunisation. It is the vaccine that was the unknown and therefore feared aspect of the problem.

The pertussis crisis also marks the point at which the field of public health began to incorporate emergent social actors in the fields of media and public relations. This crisis point was a highly mediated event, and historians and contemporary commentors alike have been critical of the role played by the press in sensationalising the harm alleged to have been caused by the vaccine. That the crisis played out before the public, through the press, the courts and in political life, leading, in both the UK and the US, to legislation to codify civil liability and compensations due to parents of vaccine-damaged children (the Vaccine Damage Payments Bill (1979) in the UK; the National Childhood Vaccine Injury Compensation Act (1986) in the US) certainly made the case more amenable to media sensationalism. However, in the UK the government also found itself behind-hand in the exploitation of media to reach parents and concerned parties. An early start in lobbying was established by the newly formed Association of Parents of Vaccine-Damaged Children (APVDC), a group set up by two mothers whose children had become disabled after they had received vaccinations for polio. The APVDC was able to benefit from the expertise of disability campaigners, in particular Jack Ashley, MP, who had a strong profile in the Labour government of Ted Heath as a supporter of disability rights. As Millward points out, the critical role played by advocacy groups like the APVDC shifted the framing of this concern:

> By basing …(their)… arguments on the existence rather than the extent of vaccine damage, the APVDC was in an advantageous position. It did not have to necessarily prove that vaccine damage was widespread, nor to concern itself with the relative weightings of the risks of vaccination versus the disease itself. All that was required was proof that the hazard of vaccine damage was real, and that the number of cases in the country was above zero. (Millward 2019: 157)

In 1974, a report was published by doctors at Great Ormond Street Hospital in the UK (Kulenkampff et al. 1974) describing some 36 cases of identifiable neurological complications subsequent to vaccination with the combined diphtheria, pertussis and tetanus vaccine (DTP). In a pattern which was to become common in subsequent medical controversies (see in particular the discussion of MMR in Conis 2015), the debate swiftly moved from the pages of specialised medical journals to the front page of UK newspapers. Public health officials, on the back foot, struggled to craft a compelling rebuttal to the poignant images of brain-damaged children which began to appear in the newspapers and, most significantly, in television documentaries. In the US, the WRC-TV documentary, *DPT: Vaccine Roulette* (Lea Thompson, writer and producer, 1982) led to the establishment of a pressure group, Dissatisfied Parents Together (later the National Vaccine Information Center (NVIC)) to campaign for federal vaccine damages payments and greater government oversight of the pharmaceutical industry. The documentary was criticised as unbalanced with little coverage given to the threat of pertussis itself. It was however influential, making compelling use of the visual medium (*Washington Post*, 28 April 1982).

Turning now to look at the reportage from the UK, we will begin with the quantitative analysis. The first point to note here is that the key themes which emerged from the reportage in the UK showed a sharp division between publications graphs. *The Daily Mail* (Fig. 5.1) had early taken the lead in promoting and developing the story of pertussis and vaccine damage. Almost all of the coverage here came from or was instigated by one writer, Health Correspondent John Stevenson, who waged a

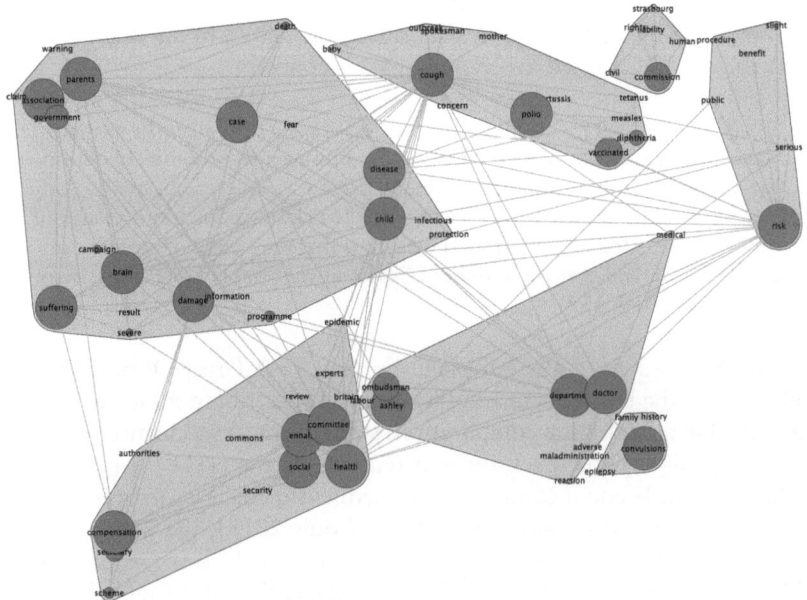

Fig. 5.1   *Daily Mail*/pertussis

long-running campaign to publicise the purported risks of the vaccine and later went on to champion the campaign for reparations for brain-injured children. The coverage therefore forms up here in four main blocks. The first includes 'doctor', 'family', 'warnings', 'health', 'child' and 'department'. The second includes 'parents', 'medical', 'association', 'risk' and 'damage'. The third includes 'baby', 'suffer', 'programme', 'government' and 'inquiry'. The final block is squarely associated with the pieces discussing the proposed Vaccine Damage Payments Bill, and includes 'minister', 'campaign', 'safety', 'committee', 'commons', 'principle', 'secretary' and 'scheme'. Notably, two of the groupings, groups two and four, have named individuals in the cluster. In the case of cluster two, this is Stevenson, the author himself, not here in the guise of writer but as actor within the text. In the case of cluster four, this is Ennals (David Ennals, minister for social services), Rosemary (Rosemary Fox, chairman of the APVDC) and Ashley (Jack Ashley, MP). This points to the extent to which *The Daily Mail* represents the pertussis vaccine controversy in combative terms, since the different actors are used to represent and dramatise the tensions of the rollout. Jack Ashley was of course a known figure in the field of health and disability policy, whilst Ennals, as secretary of state for social services led and personified the official responses. *The Daily Mail*'s coverage was almost universally focused on the campaign for compensation and recognition of the apparent danger of the vaccine, as against *The Daily Telegraph* (Fig. 5.2), which took a broader approach. In this respect, the *Mail*'s approach was that of an actor within the campaign rather than a source of information about it. The paper was a key player in the campaign to force the Callahan government to offer reparations to affected families.

It is also interesting to note the distribution of terms 'doctor' and 'medical', with 'doctor' linking to 'warning' and 'health', whereas 'medical' is located in the cluster on 'association', 'parents' and 'damage'.

The net for the third cluster ('baby', 'suffer', 'programme', 'government', 'inquiry') evidences a co-location of the two most emotionally resonant terms ('baby' and 'suffer') in the same context as the bureaucratic 'programme', 'government' and 'inquiry', underscoring the extent to which the vaccine had become an object of legal disputation. It also evidences the contrastive pairing used by *The Daily Mail* in its campaigning format, which has the effect of amplifying the emotional content of the coverage, as discussed below.

In the case of the *Telegraph*, the coverage was more distributed, forming into more diverse blocks. The terms 'government', 'disease', 'child',

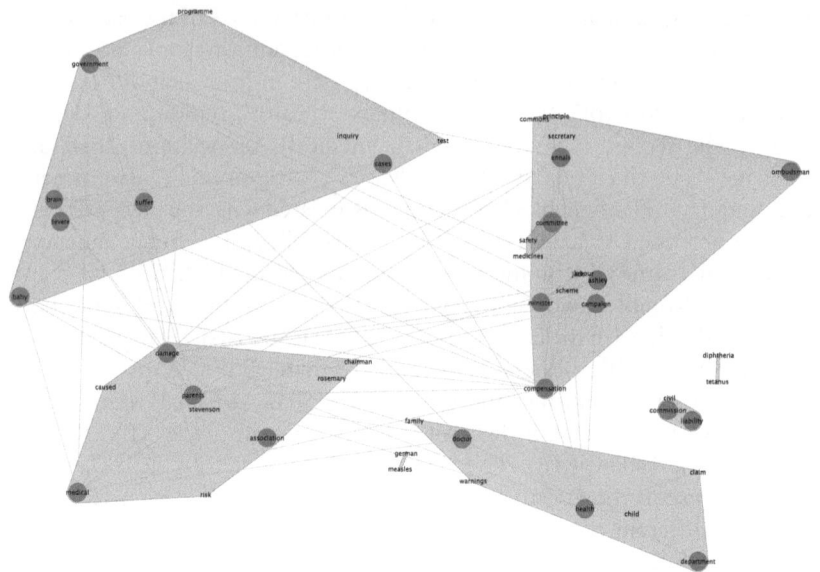

Fig. 5.2 *Daily Telegraph*/pertussis

'fear', 'suffering' and 'parents' form into one block. A second block covers 'compensation', 'security', 'Ennals', 'committee', 'experts', 'health', 'social', 'scheme' and 'authorities'. The third links 'polio', 'cough', 'diphtheria', 'tetanus' and 'measles' along with 'vaccinated', 'mother', 'outbreak' and 'concern'. The fourth is a set of 'islands' in which 'risk' links to 'serious' and 'benefits' and 'doctor' to 'ombudsman' and 'Ashley'. This latter is too distributed to be significant. A final 'island' contains terms which link to human rights and the legal implications of the UK's stance on vaccine damages compensation. These include 'Strasburg', 'human', 'commission' and 'civil'. It is interesting that these are at a remove from the remainder of the net showing the specificity of the discourse in respect of this theme.

The first cluster is largely descriptive, although it is significant that the descriptor 'parents' appears in the cluster with 'government' rather than 'compensation', and 'mother' appears separately to 'parents'. The linkage cluster around 'polio', 'diphtheria', 'tetanus' and 'measles' is likely to reflect the fears that the reduction in the uptake of the pertussis vaccine would see knock-on effects on the uptake of vaccinations more generally for children in the UK. The language of risk, as noted in the qualitative analysis (see below), is not captured by the sematic network analysis here, appearing as a much broader and more pervasive category which links into a growing focus on the role of parents in this vaccine programme. It is interesting that the descriptive cluster which includes 'experts' and 'health', in the case of the *Telegraph*, includes 'committee', 'social', 'scheme' and 'authorities', whereas for *The Daily Mail* 'health' remains tied to 'doctor/s', 'warning' and 'family'. 'Health' for *The Telegraph* then was more a question of planned social care and government strategy than of the personal relationship between doctors and parents presupposed by the adjacency of 'family' in this cluster.

Qualitative Points: Emotion versus Bureaucracy

The first point to note here is the extent to which the coverage of pertussis was characterised by a greater emotional investment than was the case with respect to polio. As Wahl-Jorgenson notes, the development of journalism as a profession has been profoundly tied to the handling of emotion in reportage. She cites Bob Franklin's oft-quoted dictum that tabloid journalism exists 'less to inform than to elicit sympathy, a collective "Oh how dreadful" – from the readership' (Franklin 1997, cited Wahl-Jorgensen 2019: 41). This argument, she continues 'creates a binary opposition between "informing" and "eliciting sympathy," whereby emotion appears as the antithesis of the dominant vision of journalism as a "fact-centered" rational discursive practice designed first and foremost to provide information' (ibid.). However, emotions can be repressed but never denied, and, as the author goes on to argue, modern journalism depends on the enactment of emotion as a strategic ritual in exactly the same way as, and as a complement to, the enactment of objectivity observed by Tuchman in her landmark study 'Objectivity as Strategic Ritual' (1972). Emotion in journalism is enacted through the bracketing off of emotional terms, descriptions of emotion on the one hand and, of greater significance to this account, drawing out a narrative which is driven by a series of emotional crisis points and comparisons on the other. Wahl-Jorgenson goes on to delineate the central role played by emotion and the skills of emotion management in journalistic content on the careers of practitioners. Drawing on Bourdieu, she argues that competence in representing and directing emotional expressions and emotive content in journalism is a key professional standard and one which forms part of the cultural capital of journalists. Using an analysis of a range of Pulitzer winning reportage, she shows how this type of expertise informs the definition of good journalistic practice in US news production.

It is a commonplace understanding that in the case of the UK the enactment of public emotion is more strongly controlled within journalism, especially in the case of the broadsheet papers, which are commonly understood to be somewhat dour and emotionally remote. However, using Wahl-Jorgeson's analysis as a sensitising concept does allow us to uncover the emotionality inherent in the coverage, albeit one which is filtered through and shaped by narrative oppositions which are more prevalent in British cultural life. One of these is the contrast observed above between bureaucracy and affect which is often subtly inflected.

Thus, to take an example from *The Daily Telegraph*'s coverage, a piece which was carried under the unambiguous headline 'MP blames vaccine for turning babies "into cabbages"', the reportage concerns the call by Jack Ashley, MP, for an investigation by the health services ombudsman of some 300 cases of children whose brain damage was alleged to have been caused by the pertussis vaccination. The narrative established draws a set of oppositions and contrasts which develops the tensions within the story. Jack Ashley is 'persistent' in his attempts to 'ventilate' the issue in the face of 'stonewalling' from government ministers. Ennals is described as having a 'direct responsibility', and Ashley is quoted as having warned of 'far reaching repercussions' and 'mass litigation'. Thus far, the only real cue to see this as an expressive and emotionally 'aware' piece of reportage is in the scare headline and particularly the reference to the children as 'cabbages'. However, we can see how the tension is developed through the contrast between the first and second parts of the story. Whilst the first part gives voice to Ashley's concerns, the piece concludes with a short rebuttal inset quoting the Department of Health's Joint Committee on Vaccination and Immunisation, which had

unanimously recommended that routine whooping cough vaccination in infancy should continue. The risk of whooping cough in infants under the age of one far outweigh the risks of vaccination. Arrangements have been made for an investigation of all children admitted to hospital with severe brain damage to determine any relationship between brain damage and vaccination. Parents need have no anxiety about their babies being vaccinated against whooping cough if such a step is recommended by their family doctor. (Loshak 1977)

The enactment of a ritual of objectivity, giving balance to 'both' sides and equal weight to arguments, actually furthers the emotional and dramatic drive of the story. In the first instance, this occurs through the presentation of the duality of 'both' parties. Ashley and the campaigners for compensation for brain-damaged children as a result of the vaccine are presented as being in conflict with specific parts of government, reasonably enough, but are further invoked as the only participants in this story, two groups going 'head-to-head' on the issue. This common feature of journalism creates, as many commentators have noted, a rivalrous 'them and us' mode of discourse which disallows mutual ground or shades of opinion. Leach and Fairhead (2007) noted this also in their review of literature concerning media roles in stoking fears of the MMR vaccine, namely 'the tendency to give undue coverage to personalized stories of alleged vaccine damage, and to "David and Goliath" stories of struggles against the scientific establishment that make good copy'. Additionally they observe that there is 'a tendency for media stories to pit two sides of a controversy against each other as if the evidence for each side were 50:50' (2007: 17).

Moreover, the depiction of the two parties is highly motivated. Ashley, the 'persistent' voice of those 'healthy' babies who have become 'cabbages', certainly benefits from the contrast with the deference to medical authority ('if such a step is recommended by their family doctor'), statistical equivocation and proposed investigation strategy (a seemingly prevaricatory follow-up investigation of the causes of brain damage in children requiring in-patient care in a hospital setting). The result then is a contrast between the energy and initiative of Ashley and the apparent cold bureaucratic cavilling and procrastination of the secretary of state.

A second aspect of the enactment of emotionality in the texts is through the attribution of and direct description of emotion. If we return to Wahl-Jorgenson's analysis, a further way which she identifies as key to 'managing affect' is the 'outsourcing' of emotion to other actors or sources within the text. Journalists use this technique of depicting, or alternatively attributing or noting the absence of, emotions in those they are describing in their coverage, which cues the reader to understand the affective elements of the story. So, to take an example, we could look at *The Daily Mail*'s ongoing coverage of the compensation campaign, carried under the headline 'Cash pledge for vaccine damage is refused. Families wait on stonewall Ennals'. Here, Ennals is described as 'angry', having 'sincere sympathy' but being 'in conflict' with the medical profession and 'deepening public anxiety'. The *Mail* was elsewhere engaged in a rather bitter attack on Ennals personally. Although their coverage of the secretary of health's stance on vaccination retained a degree of cordiality, that of Ennals' proposals to limit tobacco consumption through banning some forms of advertising was less restrained, with Ennals characterised as 'sanctimonious', 'prissy and arrogant', 'puritanical and patronising', 'the "caring" man who skilfully loses sight of his conscience on the whooping cough immunisation issue once politics becomes involved', who shows 'a fanatical inflexibility, incapable of understanding ordinary everyday life', and who allows 'desperately sick patients be dragged to operating theatres through filthy basements in hospital slums—because his dogma won't let him move an inch on finding alternative ways to finance the Health Service' ('The Sanctimonious Mr Ennals', *The Daily Mail*, 9 March 1977). Ennals was personified as 'stonewall Ennals', a monument to bureaucratic artifice, the ultimate 'baddy'. However, the *Mail* was not the only paper to depict Ennals in such damning terms. *The Spectator*, for example, carried an editorial which characterised Ennals himself as 'crass', 'brutal' and showing a 'vicious insensitivity':

> If, on the other hand, the parent decides that the vaccine should be administered, he is deliberately putting his child into a game of Russian roulette, deliberately inviting the chance of brain damage in order to guard against a vague possibility. For that dilemma Mr Ennals … showed not the slightest comprehension or sympathy. (Cosgrove 1977)

A final feature here of the way in which the bureaucracy versus emotion theme appears is the use of the personalised account, or personalised storytelling. Wahl-Jorgenson saw the 'anecdotal lead and personalized storytelling – as markers of a strategic ritual of emotionality' (2019: 46) and used stories which opened with a specific personal account as a way to identify emotionally literacy/emotional cultural capital in newspaper journalism. In the case of pertussis, it is obvious that a number of highly emotive stories would always form the basis for further coverage. The campaigning format adopted by *The Daily Mail* lent itself particularly to the use of individualised and personalised stories. Thus, for example, some of the stories took parents as the protagonist, in particular those that focused in on the burden of care, the shock of a diagnosis that seemed to come 'out of the blue'. Others spoke directly for the infant sufferers. Jack Ashley, for example, was quoted at length by *The Telegraph*:

"While I recognise the fear that my discussion of risks may impair the public's confidence in immunisation procedures. I believe it is time the conspiracy of silence was ended because too many young lives are being devastated." Mr Ashley quoted from a letter from a mother whose daughter born in 1960 was taken ill after a triple vaccine injection for diphtheria, tetanus and whooping cough. The letter said that at 13 Carolyn could not walk, talk or help herself in any way, and could not even hold her head up and was of no mental age at all. "Tragedy for a child, heartbreak for her parents," said Mr Ashley. ('Vaccine Risk to Children', *The Daily Telegraph*, 1 February 1974)

Cases such as this clearly aimed to provoke an emotional investment on the part of the readership, being colloquially 'every parent's nightmare'. Occasional personalised stories opened with a perspective from the medical profession and invited the readership to experience vicariously the conundrum faced by well-meaning practitioners (e.g. '*Whooping Cough Fear by Doctors*', *The Daily Telegraph*, 22 July 1977). Whichever figure is chosen, the personalised introduction to the pieces serves to democratise the danger. The risk is frequently described in terms which suggest a degree of randomness (see below for further discussion), and therefore the sense that this could affect anyone. Personification in this way creates not only an 'any man' but also an 'everyman' in respect of risks.

Finally, such techniques are not really an invitation to sympathy so much as a requirement of it. Few readers would feel comfortable responding to such emotive content with a rebuttal based on factual or philosophical grounds. Therefore, for the reader, only one subject position is available, namely that of sympathetic friend or confidant. Any other response requires us as readers to don the cold mask of the bureaucratic antihero. This framing then foregrounds the moral and affective and establishes distance from policy and cost/benefit frames.

Science and Risk

A key feature in the reportage is the role played by the representation of scientific expertise and technological innovation in the rollout. The depiction of scientific endeavour within the coverage on pertussis was strongly inflected by the political character of the newspapers in a way that was less true of the polio epidemic of the 1950s. Science and technology by this point had come to be far more politically and ideologically loaded in contemporary discourse. The 1960s saw science becoming a matter of popular contestation around values. As Matthew Francis argued in 2013, on the 50th anniversary of Prime Minister Wilson's 'White Heat' speech, the contestation around the value of science was always at heart one of political affiliation. Presenting Labour as the party which would advocate for science and technological expertise, as Wilson did in the 1960s, allowed him to draw a circle around the divisions within his own party by referring up to a 'greater good'. At the same time, Francis points out, this had the fortuitous additional affordance of re-situating Labour's traditional working-class constituency to focus on an increasingly skilled workforce:

> This strategy was also specifically designed to appeal to the skilled workers that were thought to be drifting away from Labour, with its emphasis on the importance of formal qualifications and technical expertise. (Francis 2013)

By the 1970s, technology took on an idealist cast in popular discourse, being seen as the only likely font of solutions to ongoing social and economic ills. Technology was seen to transcend the social problems which had dogged Britain in the seventies, with the tide of class and racial discord which came on the back of economic recession. The shift here was from the presentation of science as a moral neutral, which had made it available for co-optation by more utopian social and cultural movements, to the active investiture of technology as a source of moral good which fuelled the fervid hopes of the early cyber-utopians.

Against this background, it is hard to imagine a more emotionally loaded case than that of childhood vaccinations in the matrix of representations of expertise. As a known illness, as noted above, whooping cough carried 'domestic' connotations, evoking the world of the nursery and the solid paternalism of the 'family doctor'. This was now contrasted against a more abstract language of risk and comparison with which fewer readers, and fewer journalists, were comfortable, as witnessed by the brutal characterisation of Ennals noted above. In the US, the documentary *Vaccine Roulette* (WRC-TV Washington, 1982) and Coulter and Fisher's popular work *A Shot in the Dark* (1991) characterised vaccines in the language of random chance. This representation was of course relative as it played out in the UK press. *The Daily Telegraph* was a great deal more comfortable with the language of risk than *The Daily Mail*. *The Daily Telegraph* used a greater number of official sources and offered less editorial framing of them than *The Daily Mail*, whose campaigning format required it to maintain a distance from official narratives. Since the decision to advocate for the pertussis vaccine by health officials was most often accompanied by some assessment of the comparison of risks between whooping cough itself and the vaccine, this language was more often featured in case of the *Telegraph* than the *Mail*.

On the other hand, the dissimilarities could easily be overstated. The coverage from both newspapers, while divergent in respect of content, developed structural similarities. Both papers included detail on the poor outcomes as they pertained to the vaccinations directly but turned away from depictions of the more well-known risks associated with whooping cough. Thus, although the language of risk and measurement of alternative probabilities features in the reporting, the structure of the reportage does not follow that, with a clear bias towards the representation of novel threats.

This is of course, as widely noted, a common feature of risk discourse more generally. Unknown risks and new threats tend to be exaggerated, or at least overemphasised, in popular concerns, and this is as much the case in respect of health as it is in any other area of public concern. Indeed, in the case of vaccine side effects, this is more pronounced than would be otherwise the case, and this can be accounted for in the comparison between vaccination concerns. In the case of prior immunisation programmes, fears had resolved down to the probability of contracting the virus that immunisation was intended to avert. In the case of pertussis though, the side effects were not directly related to the virus but to the vaccine. For this reason, concerns around whooping cough show a greater degree of variability, and their advocates were often more sceptical than apprehensive as such. This set of vaccine concerns were more easily co-opted into broader, more politically loaded, concerns around science and democracy.

Public Relations Emerging into the Field of Health

The case of pertussis constitutes a key moment when the field of public relations began to take a role in that of health. Gangarosa et al. (1998), in their study of the impact of anti-vaccination movements on epidemic pertussis, divide countries where pushback to the whole-cell vaccine was experienced into two groups, the active and the passive. Passive groups feature more in countries where healthcare providers took a lead in withholding vaccines. They are, as the authors argue, more frequently focused around religious concerns, and parents are seldom featured in coverage and discussion. The reverse is true of active groups, where 'well-organised movements sought to stop (whole-cell vaccine use) by means of news stories, television interviews, lectures, popular articles, books, and other writings. Distraught parents whose children suffered adverse effects blamed on whole-cell pertussis vaccination featured prominently. Some outspoken medical authorities became leaders in these movements' (Gangarosa et al. 1998: 356). Both types, the authors found, involved alternative practitioners in key roles.

The story of anti-vaccine movements, for Gangarosa et al., as Millward has also argued in a different context, is 'a cautionary tale of the risks of allowing fear of vaccine safety to grow amongst the public, as well as evidence for the efficacy of mass vaccination programmes' (Millward 2016: 431). The authors use comparative data from the two groups of countries to demonstrate that the incidence of pertussis rose and fell with the waxing and waning of anti-vaccine movements. The correlation is somewhat complicated by the tendency of fears around vaccines to reassert themselves once high vaccine rates and 'herd immunity' have been established, leading to a '"tragedy of the commons"- a loss of confidence in vaccine and a resurgence of disease' (Gangarosa et al. 1998: 360).

However, as Millward goes on to argue, this is only one possible framing of the question of pertussis and anti-vaccination, and the case of the UK was to some extent unique. Thus, he argues, studies such as these

> locate the controversy almost exclusively within public health practice and policy. This ignores much of the contemporary political climate, including the growing disability movement, sweeping reforms to social security benefits for disabled people, the legacy of the recent thalidomide scandal and a deepening financial crisis. As a result, the scare has been somewhat dehistoricised... (Moreover)... these forms of analysis gloss over the political and social context of the period, and take as granted the hindsight that the pertussis vaccination was declared safe in the early 1980s. Further, they tend to disregard the historical importance of the scare in its own right in favour of wider practical questions about public health and vaccination. For public health practitioners, the scandal represents 'bad science', to borrow a term, in a world that is more prone to focus on the 'lesson of history' for concrete action rather than to understand the motives of policy actors within their own specific historical context. (Millward 2016: 431)

The case of the UK, argues Millwood, was unique insofar as it occurred at a time of transformation in disability politics and became strongly articulated with these concerns. At the time of the pertussis crisis, UK governments had started to reconsider the nature of disability, reframing it as both a medical and a social question. Campaigners successfully sought to challenge the narrative that disabled people were so by dint of physical incapacities, a framing which had previously confined 'disability' as a category to the sphere of medical science. Campaigners instead focused on 'social' disability, the role played in the perpetuation of disabilities by societal failure of responsibility to ensure equality of access for disabled as well as able-bodied people. In foregrounding this concern, campaigners and voluntary organisations drew upon the successes of prior campaigns, especially those around poverty in the UK. Voluntary groups focused on specific illnesses and incapacities, for example MENCAP (mental disabilities), and the (then named) Spastics Society (now SCOPE), which focused on sufferers from cerebral palsy. As Millward goes on to argue, these campaigns were proving successful to the extent that they were able to separate a moral argument from a scientific one. In the case of the campaign for government compensation, the vaccine damages campaign, the moral argument was an early victory, leaving aside questions of causality to be deliberated in other spheres. Millward cites a timely piece published anonymously in the *British Medical Journal*, which argued:

> The moral justification for compensation … is based on the social contract. National immunization programmes not only aim to protect the individual but also to protect society… If individuals are asked to accept a risk (even a very small one) partly for the benefit of society then it seems equitable that society should compensate the victims of occasional unlucky mishaps. (Cited, Millward 2016: 437)

The idea of an 'occasional' mishap was central. The later agreement to compensate individuals showed government confidence that the vaccines were safe and that only a small number of cases could be expected. (Millward 2016; see also Millward 2019).

The APVDC was successful, it can be argued, to the extent that they were able to sidestep questions of causality and scope whilst reasserting the primacy of the moral position. As Millward concludes, the APVDC used techniques which were becoming increasingly embedded in the toolboxes of voluntary organisations and pressure groups. The APVDC used professional research, gathering support from medical experts, in this case Prof Gordon Stewart and Dr John Wilson, amongst others. They were also well organised, establishing key connections with strong allies, beginning with journalists such as Celia Hall (then Health Correspondent for the *Birmingham Post*, later medical editor for *The Telegraph*), MPs Barbara Castle and Jack Ashley, who brought parliamentary voices to the proceedings and helped the nascent group to clarify their aims and resources (see Fox 2016). The Association's use of the media, including local and national radio as well as print journalism, was sophisticated and sustained. It was also effective in gathering and collating instances and personal stories to counter, rhetorically at least, the sense of the problem of vaccine damages as an idiosyncratic and isolated phenomenon. Thus, the vaccine damages campaigners had a complex relationship with the question of scale. On the one hand, they needed to demonstrate the plurality of cases to warrant state consideration; on the other, they also needed to not overstate this in order to obtain it. It is interesting for our purposes, however, to note the bracketing of moral and scientific questions from each other and the way in which this comes to be central to the resolution of the crisis. I will return to this issue below.

PATIENT ADVOCACY

As Alex Mold (2015) has argued one of the key changes in healthcare in the second half of the twentieth century was the changing relationship established between patients and medical providers. This began to shift towards the incorporation of a model of the patient consumer as an actor within the system. Mold argues that whilst to begin with the idea of the patient consumer was inimical to the values of the newly established National Health Service (NHS), and therefore strongly resisted, by the 1970s greater consumer representation was featuring more in healthcare planning, through community health councils (Mold 2015: 1). However, it would not be until the rise of free market Conservatism in the 1980s under the leadership of PM Margaret Thatcher that the notion of the 'citizen consumer' and 'patient consumer' became more prevalent. When this occurs, as Mold further argues, the 'conceptualisation of patient consumerism revolves around markets and choice, not rights and representation' (ibid.: 2). This, he argues, built the system of patient representation around a set of discourses concerning rights as mediated through consumption, and further embeds into healthcare: firstly, a legitimisation of

inequity of provision, following the model of consumerism more generally, and secondly, a division of resource based on the 'cultural capital'
available to the individual to be an effective consumer. For working-class
patients, whose upbringing had socialised them into a sense of medical
noblesse oblige and gratitude towards medical professionals, the ability to
challenge and advocate for oneself on consumer-based grounds was quite
alien. Mold cites Trentmann's analysis of the extent to which the ability to
mobilise consumer discourses into coherent demands depends on the cul-
tivation of 'political synapses', which allowed individuals to parse their
experiences into a language of rights and entitlement (Mold 2015: 4).
Finally, Mold points to the way that the model of the patient consumer
follows and replicates a wider change in the definition of consumerism
which bifurcates that of the 'citizen consumer' and the 'rational consumer'
along political lines. Thus, the rational consumer is one whose mastery of
relevant information enables them to extend their agency and gain individual benefits, whilst the figure of the citizen consumer, so foundational
to the aspirations towards a 'third way' in the 1990s, mediates between
state control and the perils of a free market. Mold identifies a typography
of seven themes which animated the progress of the patient consumer
identity, and these broadly map onto sequential historical periods of the
NHS. These are 'autonomy', 'representation', 'complaint', 'rights', 'information', 'voice' and 'choice'. The different modes of engagement with
healthcare services are characteristic of the way new consumer groups
worked to provide for patient engagement with medical services.

In the case of pertussis, the idea of choice comes very much to the fore
in the coverage, especially that of *The Daily Mail* in the campaign for reparations for parents of vaccine-damaged children. Much prominence is
given to the lack of availability of information for parents on the potential
for side effects. Thus, in a piece on victim Tanya Price, the *Mail* focuses on
the parents' experiences:

> Tanya's father, bricklayer Mr Terry Price. 29, said: 'After the vaccination she
> suddenly changed from a normal healthy baby into the vegetable-like state
> she is in now,' 'Ever since then I have been trying to get hold of her medical
> records to find out exactly what went wrong, but no one will let me get my
> hands on them.' Tomorrow … (the parents will) take her to see a leading
> neurologist in Paris: Friends helped pay the fare and Mr Price did a spon-
> sored walk to raise more cash. 'It's our last hope', said Mrs Price who is six

months pregnant. 'Nobody in this country can tell us what is wrong with our child.' (Cliff 1976)

A later piece carried under the headline, 'The Tragedy of Tanya', centres on the parents' lack of informed consent in the case:

'We were never told that we had a free choice about, it,' says her father, bricklayer Mr Terry Price... 'Now we have been informed by a specialist that nothing can be done for our child anywhere in the world. It is absolutely heart breaking.' (*The Daily Mail*, 1 June 1977)

The Prices were presented as working-class 'every men' whose efforts to both understand and ameliorate their daughter's condition were founded in working-class cultural practices. Mr Price engaged in a sponsored walk to raise funds for his daughter to be seen by experts in Paris and ran a fundraiser through his community, for example. The bewilderment of Tanya Price's parents is persistently underscored in the coverage. The paper thus takes on the role of advocating for those whose ability to campaign on their own behalf is so compromised.

Of course, it is significant here that the demand made by the parents was not for reparation, so much as information. In comparison with the Price's modest demands, the medical profession appears as secretive and evasive. The request for information and to understand what happened to their daughter fits with the model of 'citizen-consumer', advocating for equal access to information about their daughter's care. The response from the secretary of state, a reiteration of information around the relative risks, reiterates a state-based/centralised model. In this context, Kitta (2012) cites Powell and Leiss' distinction between scientific and statistical versus intuitively grounded languages of risk (Powell and Leiss 1997, cited Kitta 2012: 279). The problem of conceptualising and communicating risk cuts to the heart of that of generating trust in vaccines she argues. The problem is that the language of risk is probabilistic and the public's trust in the medical profession is founded on absolutes. Patients expect their doctors to have 'all the answers' and see assessments of risk as covering ignorance.

the ambiguity of the "expert" assessment of risk, which is a necessity when dealing with scientific matters, versus the public assessment of risk, which places a value on the individual and wants definite answers, are at odds with each other. Medical science cannot provide the public with definite answers, because it is often unaware of long-term effects and various adverse reactions. The public, however, does not understand how the medical community can be stoic in the face of death, or how it perceives any death as an acceptable risk. (Kitta 2012: 280)

The NHS and new health authorities were therefore presented as remote and patrician. With few modes of accountability to patients and enjoying a rich cultural legacy of deference to medical knowledge, doctors and medical professionals were wrong-footed by a public demand for involvement and information, founding their responses on an assumption of trust that was falling from beneath their feet.

The Campaigning Newspaper

One of the critical distinctions between the coverage of pertussis and that of polio in this sample is the presence, from *The Daily Mail*, of a large body of editorials devoted to the topic. The *Mail* took a key interest in the question of reparations required by parents of vaccine-damaged children, and throughout this period ran a campaigning/advocacy-style narrative helmed, as noted above, by the health correspondent John Stevenson. Of course, their campaign was not unique and was sometimes overshadowed by that of *The Times/The Sunday Times*. *The Times* expressed periodic outrage at the failure of the government to provide effectively for vaccine-damaged children. The 'official treatment of these tragic children is nothing short of a scandal' opined *The Sunday Times* in an editorial of 14 July 1985. 'The facts are a disgrace for a civilised nation to tolerate…These children, after all, have paid the price for the protection of the community as a whole'. In 1977, *The Times* had already laid out a similar case for compensation.

> In principle it is fair that when the Government advocates a course of action which is known to cause tragic damage in a few individually unpredictable cases, but benefits the community in general, it should accept a special responsibility towards those who suffer. This is especially so with whooping cough vaccination where the children who take the risk are already past the age at which the disease would be likely to do them much harm. (Not Really Reassuring: Editorial, *The Times*, 15 June 1977)

Advocacy journalism of course was not new at the time. We can trace the roots of this back to the rise of new journalism in the latter part of the nineteenth century (Weiner 1988). The social investigator and journalist William Stead, when editor of the *Pall Mall Gazette* called for the virtues of this kind of advocacy to be recognised within journalistic circles, described his vision of 'government by journalism' (Stead 1888). Successful campaigns run by newspapers through the end of the nineteenth century going into the twentieth century focused on a wide variety of issues, though the most prevalent were those associated with either moral questions, or questions directly pertaining to safety, particularly in the context of either work environments or consumption practices. Thus, we see campaigns both local and national examining issues as diverse as female worker safety (Malone 2002) adulteration of foods and, of course, perennial questions around morality which drew upon Stead's original campaigning for an increase in the age of sexual consent in the UK (Walkowitz 1992).

Advocacy journalism is defined by its rejection of the ethics of objectivity and equality of access as central to reportage. Campaigning journalism involves a sustained engagement with a single issue by a newspaper, is usually focused on a defined outcome (in this case, provision of adequate reparations to parents for vaccine-damaged children), and features an asymmetry of sources, tending to privilege pressure groups, especially nascent ones, over official sources, and featuring their early involvement in the development of the campaign.

This often means that advocacy pieces will feature official sources in fewer contexts and in a less-sympathetic light (see Thomas 2018). It is also the case that advocacy journalism enrols a wider community of commentators, which has the effect, independently of the content and reception of the reportage, of enlarging and diversifying the range of perspectives and 'stakeholders' involved. For this reason, as Birks (2010) has argued, campaigning journalism is a key form used within left-leaning and provincial newspapers, the former as a consequence of the inevitable clash between official sources and pressure groups, the latter potentially as a consequence of the reliance of local journalists on single-issue-based, small-scale pressure groups within local communities. For this reason, advocacy journalism, when escalated to the national press, tends to intensify and laud the individual and anecdotal in presenting an issue.

These trends are strongly in evidence in the coverage of pertussis by the *Mail*. Early pieces took a strongly oppositional tone, questioning, 'Who is to blame for the great vaccination fiasco? How safe are vaccines for your child?' and committing to 'the campaign which the Daily Mail will continue to wage until the Government concedes to the need for damages paid to the parents of these children'. As health correspondent, Stevenson styles himself as the voice of damaged children, and as an active participant in keeping pressure on parliament to account for opposition to, or delays in the implementation of, the Vaccine Damages Bill. Stevenson characteristically used the personalised format in his copy, making individual children afflicted into household names. In so doing, he was materially assisted by his access to, and sympathetic presentation of, the campaigning MP Jack Ashley, whose prior work had included lobbying for the victims of Thalidomide, a health scandal which at this time was both recent and raw in the public memory.

The *Mail* was keen to eschew the image of the anti-vaccinator as a figure of fun, offering an account of

> an old family friend had an obsession about vaccination being dangerous. He'd go on and on about it day in, day out, driving everybody mad. 'Oh, he's a crank,' people would say. 'He's got a bee in his bonnet.' But he also had, as things have turned out, some pretty accurate knowledge.

The piece contrasts the 'insider' knowledge attributed to the 'old family friend' against an informational vacuum on the part of the health authorities, who are not specified in this piece.

> We are not told that there are possibly hundreds of teenagers with the body of an adult and the mind of a child because they were vaccinated. They were normal healthy babies, now shuffling around playing with counting bricks, and denied compensation of any kind. Try to get any statistics and you'll be made to feel an hysteric or shuffled out as a troublemaker, but children have been vaccinated under an unthinking, unquestioning conveyor – belt system for too long. It's time parents had the facts. (Lee-Potter 1974)

The essential elements of this account then are the extent to which anecdotal evidence is made to stand as a refutation of scientific evidence; the complex playing with different types of 'knowledge' within the piece; and the extent to which an environment of doubt and ambiguity is created around the issue. The afflicted are described as 'possibly hundreds', and the absence of hard facts is raised to a virtue by the assertion that any attempt to get statistics would be resisted. The contrast between the wise insider who is made to feel 'like an hysteric' and the stonewalling officials makes refutation on the basis of scientific evidence impossible. No balance is needed in reportage, for none could be, by this formula, seen as credible. Indeed, the absence of factual information confirms the status of the journalist as part of the community they are advocating for, as an insider or 'one of 'us'. In this way, anti-vaccination discourse constrains the range of subject positions available such that the mode of persuasion/proof affiliated with 'resistant / questioning parent' is anecdotal rather than statistical, and the alternative available positions – troublemaker, hysteric – are disavowed. It is also notable that the modalities of knowledge in this account match closely those of the 'classic' conspiracy theory, as identified by DiRusso and Stansberry (2022).

The format adopted by the campaigning newspaper is an example of the ways in which transfers occur between the reified and the consensual universes of discourse (Moscovici 2000; see Chap. 2). As has been noted elsewhere, we can see that there are parallels between the structure of anti-vaccination discourses from the various points of view of policy/academic/practitioner stakeholders and the vaccine hesitant. Thus, for example, both rely on monocausal explanations, which serves to smooth the transfer of ideas across contexts. For policymakers, the fears that vaccine hesitancy would take hold focus on the role of the media in leading the public. For example, MacKinnon argued:

> The decline in the acceptance of whooping cough vaccine in Dudley has been shown to be related to publication of the reputed hazards of immunization in the mass media and, should current studies confirm that whooping cough immunization is effective and safe and should continue to be routinely offered to infants, the presenters of health topics and items in the media must surely give thought to the results, in terms of mortality and morbidity, of the ways in which such topics are presented. (McKinnon 1978: 202)

On the other hand, the fear equally expressed by the vaccine hesitant is that the media are in some ways 'in cahoots' with policymakers and the pharmaceutical industry. The appearance of a story about vaccine hesitancy is therefore sure to prove one or other of these points. The case of pertussis then offers us a study in the development and embedding of a structure of opposition within discourses of anti-vaccination. In the final chapter, I will move the story forward by a leap to consider the applicability of this to the case of Covid-19.

REFERENCES

Baker, J.P. (2003) 'The pertussis vaccine controversy in Great Britain, 1974-1986.' Vaccine, 21(25–26), pp. 4003–4010.

Beck, U. (1992) Risk Society: Towards a New Modernity. London: Sage.

Birks, J. (2010) 'THE DEMOCRATIC ROLE OF CAMPAIGN JOURNALISM', Journalism Practice, 4(2), pp. 208–223. Available at: https://doi.org/10.1080/17512780903407437.

Cliff, P. (1976) 'The baby ill after a "jab"', Daily Mail, 29 November, p. 11.

Conis, E. (2015) VACCINE NATION: AMERICA'S CHANGING RELATIONSHIP WITH IMMUNIZATION. Chicago/London: University of Chicago Press.

Cosgrove, P. (1977) 'Suffering and little children', The Spectator, 19 February, p. 13.

Coulter, H and Fisher, B (1991) DPT: A shot in the dark US: Avery Publishing Group.

DiRusso, C. and Stansberry, K. (2022) 'Unvaxxed: A Cultural Study of the Online Anti-Vaccination Movement', Qualitative Health Research, 32(2), pp. 317–329. Available at: https://doi.org/10.1177/10497323211056050.

Fox, R. (2016) Helen's Story. London: John Blake.

Francis, Matthew (2013) 'Harold Wilson's 'white heat of technology' speech 50 years on' The Guardian (UK) 19 September.

Gangarosa, E. J., Galazka, A. M., Wolfe, C. R., Phillips, L. M., Gangarosa, R. E., Miller, E., & Chen, R. T. (1998). Impact of anti-vaccine movements on pertussis control: the untold story. Lancet (London, England), 351(9099), 356–361, Lancet, 351(9009), pp. 356–361.

Kitta, A. (2012) Vaccinations and Public Concerns in History: Legend, Rumor and Risk Perception. New York: Routledge.

Kulenkampff M, Schwartzman JS, Wilson J. Neurological complications of pertussis inoculation. Arch Dis Child. 1974;49(1):46–49.

Leach, J. and Fairhead, M. (2007) Vaccine Anxieties: Global Science, Child Health and Society. London and New York: Routledge.

Lee-Potter, L. (1974) 'The questions we Must ask', Daily Mail, 25 September, p. 7.

Loshak, D., Health Services Correspondent (1977) 'Whooping Cough Fear by Doctors', The Daily Telegraph, 22 July, p. 5.

Malone, Carolyn (2002) 'Campaigning Journalism: The Clarion, The Daily Citizen, and the Protection of Women Workers, 1898-1912', Labour History Review, 67(3), pp. 281–297. Available at: https://doi.org/10.3828/lhr.67.3.281.

McKinnon, J.A. (1978) 'The impact of the Media on whooping cough immunization', Health Education Journal, 37(3), pp. 198–202. Available at: https://doi.org/10.1177/001789697803700307.

Millward, G. (2016) 'A disability act? The Vaccine Damage Payments Act 1979 and the British government's response to the pertussis vaccine scare Social History of Medicine 30:2 429–47.

Millward, G. (2019) Vaccinating Britain: Mass vaccination and the public since the Second World War. Manchester: Manchester University Press.

Mold, A. (2015) Making the patient-consumer Patient organisations and health consumerism in Britain. Manchester, UK: Manchester University Press.

Moscovici, S. (2000) Social Representations: Explorations in Social Psychology. Edited by G. Duveen. Cambridge: Polity.

Stead, W. (1888) 'Government by journalism', Contemporary Review, (49), pp. 653–657.

Thomas, R. (2018) Advocacy Journalism in T.P. Vos (ed) Journalism Boston: Walter de Gruyter 391–414.

Tuchman, G. (1972) 'Objectivity as Strategic Ritual: An Examination of Newsmen's Notions of Objectivity', American Journal of Sociology, 77(4), pp. 660–679.

Wahl-Jorgensen, K. (2019) Emotions, Media and Politics. Cambridge: Polity.

Walkowitz, J. (1992) City of Dreadful Delight: Narratives of Sexual Danger in Late-Victorian London. Chicago/London: University of Chicago Press.

Weiner, J. (ed.) (1988) Papers for the Millions: the New Journalism in Britain, 1850s to 1914. Westport, Conn.: Greenwood Press.

CHAPTER 6

Covid Discourses, Populist and Academic

Abstract This chapter reflects on the implications of previous health messaging around vaccination and inoculation for the roll-out of the vaccination programme tackling the Covid-19/coronavirus in the UK in
2020–2021. The chapter conducts a thematic analysis of reportage around
vaccination and anti-vaccination in the UK press during this period. The
chapter examines a range of key themes around the status of vaccination as
an 'icon of modernity', the nature of hesitancy in a time of scarcity and the
notion of citizen responsibilities to the state in the matter of health in
understanding the case of COVID. The significance of contested representations around mask wearing and 'sunflower' lanyards is also discussed.

Keywords Covid-19 • UK newspapers • Semantic network analysis •
Anti-vaccination • Metaphors • Discourse analysis

As noted above, the question of vaccine hesitancy has been made all the
more urgent in recent times by the experience of the Covid-19 pandemic.
This is not 'merely', though the term 'merely' would be doing a job of
work here, that the pandemic was so virulent, so wide in its reach, so
sweeping in its coverage. It is also that as a 'novel' virus, Covid lacked an
established narrative, either in popular culture or in medical and health
policy, as well as even the basis for an effective vaccine. Modes of

treatment effective in other coronaviruses were less so in the case of Covid. The virus seemed to selectively target particular groups, largely the elderly, but could have widely disparate outcomes in all groups. Some seemed to shrug the virus off with few or even no symptoms; others were made very unwell indeed. Young otherwise healthy sufferers died, with their doctors no less confused that the bereaved. Some recovered only to go on to suffer rapid onset complications within days or weeks, often leading to further mortality. Recovery times varied, with some infected and symptomatic up and about within days; for others, a slow and disjointed recovery awaited. For yet others, now estimated in the UK as 1.9 million (ONS: March 2023), the shadow of Long Covid stretched out, with neither clear cause nor cure. Terrified journalists reported weekly figures, whilst the fear that hospitals would be overwhelmed was always present. For key workers, especially within the NHS, the experience of the pandemic was crushing, as medical staff worked on, and on. Early on in lockdown, supermarkets and shops introduced capacity rationing to ensure shopper safety, but also to avoid further instances of panic stockpiling which had been seen briefly in the first days. Short-lived product runs on items not usually warehoused in large quantities, for example hand sanitiser, made for an increased sense of peril. Residential streets bristled with delivery vans and the furloughed, out for their daily exercise allowance, whilst bored children stared out from behind rainbow-painted windows. As traffic slowed, and skies cleared, the sense of time held in abeyance, a slow suspension, festered into impatience. Only a vaccine, or the brutal operation of 'herd immunity', whose key premises governments squeamishly or perhaps prudently omitted from public communications, could release a population trapped in social 'bubbles'.

Under such circumstances the pressure on vaccine researchers to come up with the goods was, to say the least, intense. The WHO quickly organised to establish principles for the distribution of vaccine globally once this had been developed, but in the absence of a viable product this was largely principled throat clearing. Scientists cautiously attempted to manage public expectations, knowing that the usual timelines for development of new vaccines would be measured in years not months. The country nevertheless held its breath.

Vaccines as 'Icons of Modernity'

As Dew argues, vaccines, 'along with antibiotics, symbolically represent the medical profession's ability to battle against ...germs' (1999: 383). Drawing on Osbourne, he contends that 'vaccination levels become important indicators of state success, that is, if vaccination levels are high the state has achieved its target, no matter what relationship this has to actual levels of health or protection from disease' (ibid.: 392). As such, vaccine figures come to stand proxy for efficiency and statecraft in a period in which the state is held to be responsible for citizen health but unable to necessarily secure it. This valuation of vaccines as part of a national discourse in the context of Covid also becomes wrapped up with the increasing suspicion of and competition between nations. As noted above, the pandemic came at a period in which international cooperation and a sense of global commonality were already failing within political discourse in many nations. The ideological overloading of vaccine administration even prior to the pandemic is therefore another element of this competition. Of course, competition and posturing over the state's responsibility in vaccine administration paled in comparison to the international investment in the production of the vaccine. Although behind the scenes cooperation was much more the case between apparently rivalrous concerns, as far as the public were concerned the race for a vaccine was tied up with nationalistic concerns.

A headline from *The Daily Telegraph*, for example, claimed that the 'Search for a vaccine is this century's Space Race – and the stakes could not be higher'

> Across the world, scientists in more than 100 disparate teams are working at breakneck speed to produce a vaccine for the coronavirus. On the table are not just millions of lives and many billions of dollars and yuan, but perhaps a century's worth of geopolitical power and status. If science is the exit to the Covid-19 lockdown, then a working vaccine holds the most likely key to the door. (Nuki 2020)

This latter point was perhaps more resonant for the reading public. Although supposed 'herd immunity' was believed to be building in the population, even in the face of the rapidly mutating coronavirus, only a vaccine would allow an end to the restrictions on liberty under which most chafed. Papers warned that the 'UK could face indefinite lockdown without a vaccine against coronavirus' (Lintern 2020). This was swiftly qualified by hopes that a loosening of restrictions might sweeten the bitter pill, whilst experts warned against both fast-tracking an inadequately tested vaccine for fear of loss of public confidence and ignoring concerns about the long-term impacts of prolonged lockdown on the population's health, including availability of routine vaccination.

It was therefore quite understandable that the news that a vaccine had been developed was greeted with enthusiasm and relief. However, although the press had presented the race for a vaccine in terms of scientific discovery, the production of the vaccine was a more varied process. As Prof Dame Sarah Gilbert, Covid pioneer, explained:

> People often think of vaccine development as being all about immunology, but we need to think about the manufacturing side as well. It was very important that we create a safe and effective vaccine. We had to make it in very large quantities for a low price. Not only that, we had to ensure it could be stored in the fridge to be used in a wide range of global health situations. (UKRI 2021)

However, for the press the vaccine breakthrough was represented in the language and through the narrative of the hero's quest. The US-based *Time* magazine, for example, spotlighted the work of the virologists who developed the mRNA platform, under the headline 2021's 'Heroes of the Year'. Whilst acknowledging that

> (t)he four were hardly alone in those efforts … Progress flows from the gradual accretion of knowledge. In the case of the COVID-19 vaccines, it started with the initially painstaking process of decoding the genomes of all living things; then folded in the development of sequencing machines that reduced that genetic reading time to hours; and finally weaved in the insights—"Put it in a fat bubble!"—that seemed to come in brilliant flashes but were actually the result of wisdom developed over decades.

Time nevertheless reasserted the individual hero narrative, beginning with the scientists' biographical 'origin stories', emphasising humble beginnings, career setbacks, acknowledging the intervention of mentors and employers who believed in their work, their singular dedication and long-standing hard work. More generally, these stories then would generally move onto the 'Eureka' moments experienced by the scientists, breakthroughs that would change the future of virology:

Thanks to the scientists leading the groundbreaking development and elegant construction of these COVID-19 vaccines, we now have a list of near-infinite possibilities. The vaccines work with a magnificence that only highlights how far science has come—and how far behind society remains in recognizing and accepting what is now possible. Our communications, our politics, our splintered cultures are still snarled in confusion and skepticism, keeping people from getting the shots. Through the harrowing first winter of COVID-19, scientists gifted humanity with the ultimate prize—a weapon to fight the pandemic. It's now up to humanity to return the favor. (Park and Ducharme 2021)

This piece, whilst in no way to denigrate the extraordinary achievements of the scientists concerned, is a good example of the construction of a scientific account using the narrative of the 'Hero's Journey', a term coined by Campbell in his 1949 work to describe a 'monomyth' of the hero. The narrative can be outlined in these terms:

A hero ventures forth from the world of common day into a region of supernatural wonder: fabulous forces are there encountered, and a decisive victory is won: the hero comes back from this mysterious adventure with the power to bestow boons on his fellow man. (1949: 28)

The turning points and themes within these three broad sections were then divided by Campbell into 17 distinct moments including 'The Call to Adventure', 'The Crossing of the First Threshold', 'The Road of Trials', 'Rescue from Without', 'Master of Two Worlds' and 'Freedom to Live'. Each of these elements included characteristic symbolism, a progress of events logical to the element and a resolution. Thus, for example, in the 'Road of Trials' section, the 'original departure into the land of trials represented (is) only the beginning of the long and really perilous path of initiatory conquests and moments of illumination. Dragons have now to be slain and surprising barriers passed—again, again, and again. Meanwhile there will be a multitude of preliminary victories, unretainable ecstasies, and momentary glimpses of the wonderful land' (Campbell 1949: 100). In the hero's journey monomyth, Campbell contends, the specifics may change but the overall structure is retained. The application of this monomyth to the stories of science, and the ways that it has been used as part of the work of public science has been widely noted (see Heylighen 2012; Storr 2020; Angler 2020).

That such a narrative makes science stories more compelling, more accessible, and brings to the work of the scientific community a much-needed public involvement and enthusiasm, however, overlooks the problem that where there is a hero, there must also be a villain.

Vaccine Refusal as a Social Problem

Going back to Dew's work on measles (1999), social problems are, he argues, easiest to establish where there is a strong monocausal narrative. Diffusion of responsibility and nuance of blame are kryptonite to the would-be moral entrepreneur. Where low rates of vaccine coverage are revealed in a population, often in the wake of an outbreak, the finger of blame will shift to vaccination or lack thereof. At the time of writing, immunologists believe the UK to be poised on the verge of outbreaks of both measles and whooping cough (January 2024) with a surge in cases in both the UK and the US from 2022 onwards. This is clearly linked to lowered vaccination rates. The UK's Health Security Agency (UKHSA) recently warned of '198 laboratory confirmed cases of measles, with a further 104 cases likely… with most occurring in children who were not vaccinated against measles, mumps and rubella' in the West Midlands (Iacobucci 2024). That the cases are linked to low vaccination coverage is uncontroversial. That it is a direct outcome of vaccine hesitancy however is controversial, as I will be arguing below. Yet the linkage of cause and effect is unproblematised in accounts which see the vaccine hesitant as the root of a societal problem.

In part, this can be explained by the relative passivity of media in reproducing the understandings of primary definers (Hall et al. 1978). As Dew (1999) also notes of the press in New Zealand, this reliance on medical experts comes about in part because of journalists' relative lack of confidence in such a specialised area, a situation which is likely to increase in severity the more politicised an area becomes. It also has a significant impact insofar as it tends to reproduce the vaccine refuser as a 'social problem'. The public discussion about viruses and vaccination, taking on the frames common for health policymakers and caregivers, inevitably focuses in on the day-to-day task of persuasion, and this increasingly takes the form of a narrow discussion of methods to encourage compliance and ensure adequate surveillance of outcomes. This leads in turn to three key consequences. Firstly, we can note that techniques of persuasion, when applied by the state to the citizen, are always going to have the greatest impact on those who have the fewest resources, be those financial, educational, social or cultural. Secondly, the focus of 'peril' shifts from the illness to the unvaccinated. In the case of measles, 'the illness came to be seen as threatening and therefore intervention and regulation were imposed upon the body to normalize all bodies. All bodies were to be made immune to the disease, which is caused by the offending organism. A further transition meant that the virus was no longer seen as the primary source of danger, but the individual who may transmit the disease, and therefore the individual had to be controlled (Dew 1999: 390). Finally, the effect of this shift is not only to individualise the 'problem' of illness but also to distort and obscure the social and economic bases of 'vaccine deficiency'. That people are not immunised is not necessarily a personal choice, but the 'good scientist/ bad refuser' narrative overwrites this distinction.

Just as with measles, so in this case with Covid, with some unique twists. The vaccine refuser/vaccine-hesitant individual was rendered publicly visible in the aftermath of the initial rollout of the vaccine. In the case of a virus for which immunisation programmes have been long established, the problem of patchy or low coverage is only made visible to the public when there is an outbreak of infection. It is at this point that the fact of declined rates of immunisation moves from the sphere of the reified (policymakers, health surveillance organisations, in the UK UKHSA, in the US the CDA) to that of the consensual. The problem of outbreaks is only visible when contagion and threat is already present.

However, in the Covid-19 pandemic, this was not the case. The vaccine hesitant were rendered visible in the public sphere right from the start. *The Sun*'s headline 'I've seen deaths… but won't have jab' is a good example. The piece highlighted common reasons for refusal, including feeling healthy, influence of social media, and quoted one refuser who claimed: "I don't want to be tracked by what is in my arm". A common narrative in news was that of the previously hesitant who become converts and advocates as a result of 'staring death in the face'. Thus, for example, *The Mirror* profiled a case of one man, described, endearingly for the paper, as a 'once happy-go-lucky binman' and a 'fit and healthy dad' who 'nearly died after believing vaccine lies and has begged people to get jabbed' (Retter 2022). The piece gave prominence to the fact that '(u)p to 90% of people in ICUs with Covid are unvaccinated or unboosted' and related this back to social media influence. The piece echoed the then prime minister's call for social media companies to regulate conspiracy theories online.

Others noted the problem of the vaccine refuser less as a matter of sanguine optimism and more as a statement of rebellion. One doctor quoted in *The Daily Telegraph* noted that through

> (t)alking to vaccine hesitant colleagues, it is apparent that stepping up the pressure is counterproductive: they are more likely to double down in their defiance. The mounting vindictiveness of public invective against "anti-vaxxers" provokes resentment and reinforces distrust of authorities. For some NHS staff, rejecting vaccination has become a gesture of personal resistance in response to a pandemic that has left them feeling exhausted and demoralised. (Fitzpatrick 2022)

The writer observed that as 'the severity of Covid-19 infection declines, coercive measures against the vaccine-hesitant seem to be gathering momentum' (ibid.). This was however a momentum that had been gathering for some time. In 2020, as the trials of vaccines continued, the *Telegraph*'s Miranda Levy asked, 'If vaccines are a shot at freedom, why would anyone refuse?' The piece characterised anti-vaxxers thus:

> These people are against inoculations – and particularly against vaccinating their own children. They think vaccines cause autism, that politicians are scaremongering to drum up profits for drug companies, and that Bill Gates will be using injections to plant microchips into their children.

The piece concludes with a quote from a UK medical virologist: 'If these people will take a gamble with their kids getting a potentially lethal disease like measles, why on Earth will they give two hoots about Grandma?' (Levy 2020). The vaccine hesitant were consistently described using the term 'brigade' (a lexical shorthand in the UK popular press to denote a group whom the writer believes to be obstinate and ignorant), or depicting them as selfish, as in this case.

Thus, vaccine-hesitant people are given emotional as well as physical 'wide berth' in these accounts. Those who 'repent' are allowed back into the moral fold, albeit 'with a long road ahead of them' to recovery. The penalty for a 'wrong decision', judged by this measure, must include public contrition as well as the shock and debilitating illness. Unless that is, the refuser and the vaccine deficient can be differentiated by another means, a means which can itself be symbolically reinforced, which brings us to the final theme here, that of obligation, scarcity and exemptions.

Scarcity and Hesitancy

I have noted above the way in which representations of vaccines vary along with their initial anchoring and the key role this plays in construction of audience identifications. Social representations act to 'express the way a group thinks of itself in its relationships with the objects which affect it' (Durkheim 1895, cited by Moscovici 2000: 158). In Moscovici's terms, this initial 'anchoring', an extension of a pre-existing category to include the 'new' and unfamiliar, is followed by objectification, which in social representations theory is to 'discover the iconic quality of an imprecise idea or being' (Moscovici 2000: 49). The polio vaccine (see above) was initially figured through and anchored in the world of manufacturing, terminology to describe it centring on 'firms', 'facilities', 'stock', 'distribution' and so on. This invites the reader to adopt a consumer identity, one which figures the vaccine as a product, a physical commodity like any other and therefore subject to quality control. This anchoring in the 'prototype' commodity (see Chap. 1), was then qualified to moralised consumption, in which, inflected by the supply issues being experienced and bureaucratic fumble-footing around the allocation of responsibilities between health authorities, centred the right to consume, rather than the wisdom/responsibility in so doing. The pertussis immunisation on the other hand was anchored in the universe of 'health risks' and responsibilities, couched in the language of governance, law and risk. Its distance from the 'prototype' 'health risks' was continuously rarefied through the work of lobbyists who sought to move the narrative along the continuum of avoidable and unavoidable risks. That they were able to do so is largely a function of their ability to qualify the initial anchoring to include a requirement for information to be available as a precondition of an assessment of risk. This narrowed the distance between vaccine-hesitant positions and those of the patient expert, whose demands for information and representation in healthcare were becoming more visible at this period through the efforts of other patient advocacy movements, disability campaigners and of course feminism (see Abrams 2019; Mold 2015).

With the Covid-19 vaccine, where the staged rollout led to some twists in the matter of both consumer and health-expert positions, anchoring becomes complex. As supplies of the competing vaccines became available, countries, following the WHO 'roadmap' guide to allocating vaccine supplies, offered appointments based on a hierarchy of need, usually starting with the eldest and primary care workers, moving onto the most clinically vulnerable, in some cases including groups rendered so on the basis of ethnicity or occupation, and moving down through the age groups. In addition, the roadmap specified the mechanisms by which vaccine allocations could be vired between nations to ensure parity. In the UK, this initial phase occurred between December 2020 and June 2021.

The rationing of vaccine appointments meant, for the majority of citizens, that there was a substantial period between theoretical vaccine availability and getting immunised. At the same time, government policies, reinforced by popular culture and terminology, inextricably tied proof of immunisation to freedom. 'Vaccine passports', QR code-supported track and trace apps, and vaccine requirements specific to given events and settings reinforced the relationship between release from lockdown and vaccine take-up. As Porat et al. (2021) found, the use of vaccine passports in the UK and Israel may unwittingly have sharpened oppositional attitudes to immunisation, and, where individuals felt coerced into accepting the vaccine, reduced uptake of boosters required to establish and maintain effectiveness.

Quite apart from this, however, the differential impact of lockdown on the basis of vaccine status rankled with exactly those groups whose likelihood of severe reaction to Covid was the least, and whose lives had been negatively impacted the most, the young. For teens and young adults, last in the queue for immunisation, often with careers on hold or lacking work opportunities and independent living, and developmentally predisposed to greater mental health fallout from isolation, the perception of unfairness was clear. Santavicca et al. (2022) note the increase of 'silent or vocal social tension, polarization, and radicalization' in the wake of the introduction of vaccine passports in Quebec, a tension they note which was related to a narrowing of spaces for democratic debate at the same point. The authors noted that the pragmatic division of essential and non-essential spaces and activities also sharpened tensions between groups based on age, increasing the perception of unequal citizenship. In the UK, this was further exacerbated by later attempts to loosen lockdowns by region creating the explicit 'tiered' confinements which stoked consciousness of existing regional inequalities. Tensions additionally flared as a result of mask mandates and exemption certification, especially between those who could not mask due to hidden disabilities and those angered by the non-conformism of others. Papers reported on angry confrontations in public places, customer facing staff in retail and service spaces requiring masks expressed exasperation and fear at their new and unwonted role in 'policing' mask wearing, concerns were raised that the recently introduced (2016) 'sunflower lanyard' initiative (aimed to allow acknowledgement of hidden disabilities) was being abused. One non-profit founder related:

> The most common type of person I've seen misuse the lanyard are antivaccination, anti-lockdown types, who tend to view disabled lives in the pandemic as inconvenient at best and expendable at worst. Not long ago on Instagram, I read a post that literally said: 'Just picked up my sunflower lanyard so I don't have to wear a mask, my disability: a serious case of loving freedom'.
>
> "When six out of 10 Covid deaths have been disabled people, I know this 'freedom' doesn't include us." (Roberts 2021)

The linkage made here between anti-vaccination proponents and mask refusers therefore depicted mask refusal as a visible public symbol of, at best anti-authoritarianism, at worst a more sinister anti-social agenda.

The ad hoc nature of lockdowns, mandates and vaccination eligibility during the Covid pandemic in the UK created an interlocking set of subjectivities around vaccination which differed substantially from those of anti-vaccination discourse at other periods. This was in large measure down to the reconfiguration of inequalities, including regional, structural, health inequality, financial and social disparities, around the issue of immunisation. However, this was also a function of the difference in visibility, both of the virus and of immunisation or lack of the same in public life. The latter was a transformation wrought by changing technologies around health and social surveillance, which it is beyond the remit of this study to explore, but which continue to have wide-ranging and persistent social implications.

References

Abrams, L. (2019) 'Lynn Abrams, "The Self and Self-Help: Women Pursuing Autonomy in Post-War Britain", 2019, 29, 201–21, 203.' Transactions of the Royal Historical Society, 29, pp. 201–21.

Angler, M.W. (2020) Telling Science Stories: Reporting, Crafting and Editing for Journalists and Scientists. London: Routledge.

Campbell, Joseph (1949) The Hero with a Thousand Faces Princeton, NJ: Princeton University Press.

Dew, K. (1999) 'Epidemics, Panic and Power: Representations of Measles and Measles Vaccines', Health, 3(4), pp. 379–398. Available at: https://doi.org/10.1177/136345939900300403.

Fitzpatrick, M. (2022) 'Sacking vaccine refuseniks harms the NHS; The Surgery DOCTOR'S DIARY', The Daily Telegraph (London), 24 January.

Hall, S. et al. (1978) Policing the crisis : mugging, the state, and law and order. London: Macmillan.

Heylighen, F. (2012) 'Heylighen, F., 2012. A Tale of Challenge, Adventure and Mystery: towards an agent-based unification of narrative and scientific models of behavior. Brussels, Belgium.' ECCO Working Paper Vrije Universiteit Brussel, Belgium. Available at: https://www.researchgate.net/publication/268265889_A_Tale_of_Challenge_Adventure_and_Mystery_towards_an_agent-based_unification_of_narrative_and_scientific_models_of_behavior (Accessed: 23 January 2024).

Iacobucci, G. (2024) 'Measles: Warning given over low MMR uptake after cases rise to 200 in West Midlands', BMJ, 384.

Levy, M. (2020) 'If vaccines are a shot at freedom, why would anyone refuse?; The Covid-19 trial halted last week due to an "unexplained illness" has bolstered the antiinnoculation brigade', The Daily Telegraph (London), 14 September.

Lintern, S. (2020) 'Coronavirus: UK lockdown could be indefinite until a vaccine is found, warn scientists advising government; "We are essentially waiting for a vaccine," experts say. "A vaccine is not five months away. We know it's at least 12 to 18 months away. So we will have difficult choices to make".', The Independent (United Kingdom), 16 March.

Mold, A. (2015) Making the patient-consumer Patient organisations and health consumerism in Britain. Manchester, UK: Manchester University Press.

Moscovici, S. (2000) Social Representations: Explorations in Social Psychology. Edited by G. Duveen. Cambridge: Polity.

Nuki, P. (2020) 'Search for a vaccine is this century's Space Race - and the stakes could not be higher', The Daily Telegraph (London). 11 April.

Park, A. and Ducharme, J. (2021) 'Time Heroes of the Year 2021: THE MIRACLE WORKERS', Time, 13 December.

Porat, T. et al. (2021) '"Vaccine Passports" May Backfire: Findings from a Cross-Sectional Study in the UK and Israel on Willingness to Get Vaccinated against COVID-19. 9, 902.', Vaccines [Preprint].

Retter, E. (2022) 'Jab lies so nearly killed me; CORONAVIRUS CRISIS: PATIENT'S STARK WARNING', Daily Mirror, 7 January.

Roberts, D. (2021) 'Sunflower disability symbol "hijacked" by some anti-maskers.' South Wales Echo, 5 August.

Santavicca, T., Vanier-Clément, A. and Rousseau, C. (2022) 'Preventing and Appeasing COVID-19 Vaccine Tension in Schools to Protect the Well-Being of

Children and Adolescents in Québec', International Journal of Child and Adolescent Resilience, 9(1).

Storr, W. (2020) The Science of Storytelling: Why Stories Make Us Human, and How to Tell Them Better. London: William Collins.

UKRI (2021) 'The story behind the Oxford-AstraZeneca COVID-19 vaccine success', 30 November. Available at: https://www.ukri.org/news-and-events/tackling-the-impact-of-covid-19/vaccines-and-treatments/the-story-behind-the-oxford-astrazeneca-covid-19-vaccine-success/ (Accessed: 23 January 2024).

CHAPTER 7

Conclusion

Abstract This chapter summarises findings from three case studies of newspaper representations of inoculation and vaccine hesitancy/refusal in historical contexts, that of polio in the period 1955–1960, that of pertussis in the period 1974–1979 and that of Covid-19 in the period 2020–2021. The ways in which these different vaccination 'crisis points' were framed, this chapter argues, reflects and cements different understandings of and approaches to risk, community and marginalisation on the part of individuals and policymakers. The chapter argues that risk, far from being an abstract question of probability, is formed within communities, and creates and reinforces emergent commonalties.

Keywords Risk • Anti-vaccination • Marginalisation • Newspapers • Representation • Vaccine hesitancy

This work has traced the changes in anti-vaccination discourse between two crisis points in the UK, that of polio and that of pertussis and sketched an approach to understanding how this discourse, unique or shared, offers us a way into unpicking the nature of discourses of vaccine hesitancy in recent times. In seeing anti-vaccination in instrumental terms as an obstacle, we often treat the vaccine hesitant as a unity, working on the assumption that the same or similar sets of fears and desires motivate the groups.

© The Author(s), under exclusive license to Springer Nature Switzerland AG 2024
A. Cavanagh, *Anti-Vaccination and the Media*,
https://doi.org/10.1007/978-3-031-70559-5_7

We also saw this clearly in the case of Covid in which campaigns to encourage uptake of vaccines, whilst often differentiated by target ethnic groups, still used the same structure of messaging regardless of the group addressed. Of course, the danger of treating people as a unity is that they will swiftly become so.

This work is not then intended as a historical cautionary tale, allowing us to look into the past for lessons on how to 'handle' an 'intransigent' group. Whilst as noted above in emergency historians are often 'rushed to the scene' in the hopes of a salutary 'lesson from history', it is hard to argue against Hegel's gloomy Eeyorism that history can only teach 'that nations and governments have never learned anything from history, or acted upon any lessons they might have drawn from it'.

Although it would be wonderful to be able to offer a communicative key which might unlock health equality, historians are well aware of the limitations of their craft. Rather, I want to use the example of anti-vaccine discourses to unpick what happens when people feel themselves to be unheard. The consistent theme that comes through the reportage in the two cases I am looking at, polio and whooping cough, is of people feeling that their concerns were sidelined, refused, contradicted but, most importantly, never treated with respect. I argue firstly that the politicisation of this group is illogical, though, as the comparison between the UK and the US shows not a necessary outgrowth of this marginalisation, and secondly that their politicisation is a cyclical process, one which feeds distrust and anxiety.

As noted above, in the cases of both polio and pertussis, the media, policymakers and academics invoked the language of risk, in respect of the individual in the case of pertussis and the nation in the case of polio. As Leach and Fairhead (2007) have noted in their study of vaccine anxieties, risk is a dominant frame in vaccination discourses cross-nationally. However, there is, as they go on to note, a confusion in academic and practitioner accounts in understanding risk. Firstly, they argue that such accounts embed the assumption that risk is a matter of individual decision, a 'moment of calculation by a single mind. Yet in reality, such questions are often dealt with in ongoing processes embedded in personal history, social relations and interactions, which may involve discuss' (2007: 26). Secondly, they note, risk is a game of averages:

> Generalized risk calculations are grounded in a view of a general population, and an average person (in the case of vaccination, an average child). Yet this

overlooks the possibility that people do not consider themselves and their particular child as average, and thus do not feel that these calculations could or should apply to them. (ibid.)

Moreover, they argue, risk as presented by policy stakeholders assumes a universal character as though there is a single meaning, from which all other interpretations are a distortion brought about through 'false consciousness' (ibid.). It is hard to imagine a conference of media theorists at which any such assertion would receive more than a polite hearing and some constructive feedback. The notion of meaning as either pure truth or culpable delusion is thoroughly discredited.

Instead, we must recognise that risk is formed in communities, just as any other form of signification. Stakeholders often take for granted

> a conflict of interest between rather singular visions of the individual and public good; between being selfish and public-spirited. Yet people belong to many social worlds. Important as the ongoing debate about individual versus social risks and benefits is, it overlooks the variety of collectivities and forms of common good that people may already be part of, and that shape their thinking and practice around engaging with technologies. … communities that are often based on shared knowledges and values. This raises further questions about vaccination: what kinds of solidarities are emerging, among whom, and what serves to unite them? (Leach and Fairhead 2007: 27)

REFERENCE

Leach, J. and Fairhead, M. (2007) Vaccine Anxieties: Global Science, Child Health and Society. London and New York: Routledge.

REFERENCES

Abrams, L. (2019) 'Lynn Abrams, "The Self and Self-Help: Women Pursuing Autonomy in Post-War Britain", 2019, 29, 201–21, 203.' Transactions of the Royal Historical Society, 29, pp. 201–21.

Anderson, C.W. (2018) Apostles of Certainty: Data Journalism and the Politics of Doubt. United Kingdom: Oxford University Press (Oxford Studies in Digital Politics). Available at: https://doi.org/10.1093/oso/9780190492335.001.0001.

Andrejevic, M. (2014) 'The Big Data Divide', International Journal of Communication, 8, pp. 1673–1689.

Angler, M.W. (2020) Telling Science Stories: Reporting, Crafting and Editing for Journalists and Scientists. London: Routledge.

Aupers, S. (2012) '"Trust no one": Modernization, paranoia and conspiracy culture', European Journal of Communication, 27(1), pp. 22–34. Available at: https://doi.org/10.1177/0267323111433566.

Baden, C. and Sharon, T. (2021) 'BLINDED BY THE LIES? Toward an integrated definition of conspiracy theories', Communication Theory, 31(1), pp. 82–106. Available at: https://doi.org/10.1093/ct/qtaa023.

Baker, J.P. (2003) 'The pertussis vaccine controversy in Great Britain, 1974-1986.' Vaccine, 21(25–26), pp. 4003–4010.

Barkun, M. (2015) 'Conspiracy theories as stigmatized knowledge', Diogenes, 62(3–4), pp. 114–120. Available at: https://doi.org/10.1177/0392192116669288.

© The Author(s), under exclusive licence to Springer Nature Switzerland AG 2024
A. Cavanagh, *Anti-Vaccination and the Media*,
https://doi.org/10.1007/978-3-031-70559-5

Barnes, T.J. (n.d.) 'Big data, little history', Dialogues in Human Geography, 3(3), pp. 297–302.

Beck, U. (1992) Risk Society: Towards a New Modernity. London: Sage.

Billig, M. (1987) Arguing and thinking. A rhetorical approach to social psychology. Cambridge: Cambridge University Press.

Bingham, A. (2010) '"The Digitization of Newspaper Archives: Opportunities and Challenges for Historians"', Twentieth Century British History, 21(2), pp. 225–231.

Birks, J. (2010) 'THE DEMOCRATIC ROLE OF CAMPAIGN JOURNALISM', Journalism Practice, 4(2), pp. 208–223. Available at: https://doi.org/10.1080/17512780903407437.

boyd, danah and Crawford, K. (2012) 'CRITICAL QUESTIONS FOR BIG DATA', Information, Communication & Society, 15(5), pp. 662–679. Available at: https://doi.org/10.1080/1369118X.2012.678878.

Brandes, U. and Wagner, D. (2004) 'Analysis and Visualization of Social Networks', in M. Jünger and P. Mutzel (eds) Graph Drawing Software. Berlin, Heidelberg: Springer Berlin Heidelberg, pp. 321–340. Available at: https://doi.org/10.1007/978-3-642-18638-7_15.

Brandt, A.M. and Botelho, A. (2020) 'Not a Perfect Storm – Covid-19 and the Importance of Language', New England Journal of Medicine, 382(16), pp. 1493–1495. Available at: https://doi.org/10.1056/NEJMp2005032.

Brügger, N. and Finnemann, N.O. (2013) 'The Web and Digital Humanities: Theoretical and Methodological Concerns', Journal of Broadcasting & Electronic Media, 57(1), pp. 66–80. Available at: https://doi.org/10.1080/08838151.2012.761699.

Callaghan, P. and Augoustinos, M. (2013) 'Reified versus consensual knowledge as rhetorical resources for debating climate change', Revue internationale de psychologie sociale, 26(3), pp. 11–38.

Campbell, Joseph (1949) The Hero with a Thousand Faces Princeton, NJ: Princeton University Press.

Caplan, D. (2016) 'Reassessing Obscurity: The Case for Big Data in Theatre History', Theatre Journal, 68(4), pp. 555–573.

Chopra, A. K., & Doody, G. A. (2007). Schizophrenia, an illness and a metaphor: analysis of the use of the term 'schizophrenia' in the UK national newspapers. Journal of the Royal Society of Medicine, 100(9), 423–426.

Cliff, P. (1976) 'The baby ill after a "jab"', Daily Mail, 29 November, p. 11.

Clow, B. (2001) 'Who's Afraid of Susan Sontag? or, the Myths and Metaphors of Cancer Reconsidered', Social History of Medicine, 14(2), pp. 293–312. Available at: https://doi.org/10.1093/shm/14.2.293.

Coffelt, A. and Djandji, A. (2023) 'Mutant metaphors: Frankenstein in the era of COVID-19.' Med Humanit, 49(2), pp. 272–277.

Cohen, Stanley. (1987) Folk devils & moral panics: the creation of the Mods and Rockers. [Third edition]. Oxford: Basil Blackwell.

Conis, E. (2015) VACCINE NATION: AMERICA'S CHANGING RELATIONSHIP WITH IMMUNIZATION. Chicago/London: University of Chicago Press.

Corner, J. (2019) 'Origins and transformations: histories of communication study', Media, Culture & Society, 41(5), pp. 727–737. Available at: https://doi.org/10.1177/0163443718820666.

Cosgrove, P. (1977) 'Suffering and little children', The Spectator, 19 February, p. 13.

Coulehan, J. (2003) "Metaphor and medicine: narrative in clinical practice." The Yale Journal of Biology and Medicine 76 (2003): 87–95.

Coulter, H and Fisher, B (1991) DPT: A shot in the dark US: Avery Publishing Group.

Craig, D. (2020) 'Pandemic and its metaphors: Sontag revisited in the COVID-19 era', European Journal of Cultural Studies, 23(6), pp. 1025–1032.

Criado Perez, C. (2019) Invisible Women: Data Bias in a World Designed for Me. New York: Abrams Press.

Dadari, I., Belt, R., Iyengar, A., Ray, A., Hossain, I., Ali, D., Danielsson, N. and Sodha, S.V. (n.d.) 'Achieving the IA2030 Coverage and Equity Goals through a Renewed Focus on Urban Immunization', Vaccines, 11(4), p. 809.

Daily Mail (1956a) 'A Town Bans Polio Drug for Children', 8 March, p. [1].

Daily Mail (1956b) 'Town Hall Know-All', 9 March, p. [1].

Daily Mail (1977a) 'The sanctimonious Mr David Ennals', 9 March, pp. 6–7.

Daily Mail (1977b) 'Tragedy of Tanya', 6 January, p. 2.

Daily Mail Reporter (1957) 'Polio Mother Dies without Seeing her Two-Day Son', Daily Mail, 12 August, p. 3.

Daily Telegraph Reporter (1955a) 'Britain Makes Polio Vaccine', The Daily Telegraph, 26 April, p. [1].

Daily Telegraph Reporter (1955b) 'Britain to Make Tests of U.S. Polio Vaccine', The Daily Telegraph, 21 April, p. 7.

Daily Telegraph Reporter (1955c) 'Polio Aid by Volunteers', The Daily Telegraph, 20 September, p. 7.

Daily Telegraph Reporter (1956) 'Town Accepts Polio Vaccine', The Daily Telegraph, 12 April, p. [1].

Davis, C.J. (2002) 'Contagion as Metaphor', American Literary History, 14(4), pp. 828–836.

De Cillia, R., Reisigl, M. and Wodak, R. (1999) 'The discursive construction of national identities', Discourse & Society, 10(2), pp. 149–173.

Dew, K. (1999) 'Epidemics, Panic and Power: Representations of Measles and Measles Vaccines', Health, 3(4), pp. 379–398. Available at: https://doi.org/10.1177/136345939900300403.

DiCenzo, M. (2015) 'Remediating the Past: Doing "Periodical Studies" in the Digital Era', ESC: English Studies in Canada, 41(1), pp. 19–39.

DiRusso, C. and Stansberry, K. (2022) 'Unvaxxed: A Cultural Study of the Online Anti-Vaccination Movement', Qualitative Health Research, 32(2), pp. 317–329. Available at: https://doi.org/10.1177/10497323211056050.

Doll, M.K. and Correira, J.W. (2021) 'Revisiting the 2014–15 Disneyland measles outbreak and its influence on pediatric vaccinations', Human vaccines & immunotherapeutic, 17(11), pp. 4210–4215.

Douglas, M. (1966) Purity and Danger: AN ANALYSIS OF THE CONCEPTS OF POLLUTION AND TABOO. London and New York: Routledge.

Dubé, E., Gagnon, D. and MacDonald, N.E. (2015) 'Strategies intended to address vaccine hesitancy: Review of published reviews', WHO Recommendations Regarding Vaccine Hesitancy, 33(34), pp. 4191–4203. Available at: https://doi.org/10.1016/j.vaccine.2015.04.041.

Durbach, N. (2005) Bodily Matters: The Anti-Vaccination Movement in England, 1853–1907. Durham and London: Duke University Press.

Ehrenreich, B. (2010) Smile or Die: How Positive Thinking Fooled America and the World. London: Granta.

Eisenberg, L. (1981) 'The physician as interpreter: Ascribing meaning to the illness experience.', Comprehensive Psychiatry, 22, pp. 239–248.

Erickson, A.T. (2013) 'Historical Research and the Problem of Categories: Reflections on 10,000 Digital Note Cards', in J. Dougherty and K. Nawrotzki (eds) Writing History in the Digital Age. Ann Arbor: University of Michigan Press, pp. 133–145.

Ewing, E.T., Gad, S. and Ramakrishnan, N. (2013) 'Gaining Insights into Epidemics by Mining Historical Newspapers', Computer, 46(6), pp. 68–72.

Fasce, A. et al. (2023) 'A taxonomy of anti-vaccination arguments from a systematic literature review and text modelling.', Nature human behaviour, 7(9), pp. 1462–1480.

Featherstone, J.D. et al. (2020) 'Exploring childhood vaccination themes and public opinions on Twitter: A semantic network analysis', Telematics and Informatics, 54, p. 101474. Available at: https://doi.org/10.1016/j.tele.2020.101474.

Fitzpatrick, M. (2022) 'Sacking vaccine refuseniks harms the NHS; The Surgery DOCTOR'S DIARY', The Daily Telegraph (London), 24 January.

Fox, R. (2016) Helen's Story. London: John Blake.

Francis, Matthew (2013) 'Harold Wilson's 'white heat of technology' speech 50 years on' The Guardian (UK) 19 September.

Frank, A.W. (1993) 'The rhetoric of self-change: Illness experience as narrative', The Sociological Quarterly, 34(1), pp. 39–52.

Gangarosa, E. J., Galazka, A. M., Wolfe, C. R., Phillips, L. M., Gangarosa, R. E., Miller, E., & Chen, R. T. (1998). Impact of anti-vaccine movements on pertussis control: the untold story. Lancet (London, England), 351(9099), 356–361, Lancet, 351(9009), pp. 356–361.

Garfield, E. (1979) Citation Indexing: Its Theory and Application in Science, Technology, and Humanities. New York: John Wiley.

van der Geest, S. and Whyte, S.R. (1989) 'The Charm of Medicines: Metaphors and Metonyms', Medical Anthropology Quarterly, 3(4), pp. 345–367. Available at: https://doi.org/10.1525/maq.1989.3.4.02a00030.

Goldstein, D. (2004) Once Upon a Virus: AIDS Legends and Vernacular Risk Perception. Logan, Utah: Utah State University Press. Logan, Utah, US: Utah State University Press.

Gould, T. (1995) A Summer Plaque: Polio & its Survivors: New Haven and London: Yale University Press.

Graham, S., Milligan, I. and Weingart, S. (2016) Exploring big historical data: the historian's macroscope. London: Imperial College Press.

Hall, S. et al. (1978) Policing the crisis : mugging, the state, and law and order. London: Macmillan.

Hamilton, S. (2010) 'Reading and the Popular Critique of Science in the Victorian Anti-Vivisection Press: Frances Power Cobbe's Writing for the Victoria Street Society', Victorian Review, 36(2), pp. 66–79.

Hanne, M., Hawken, S. and Sontag, S. (2007) 'Metaphors for illness in contemporary media.' Medical Humanities, 33, pp. 93–99.

Haraway, D.J. (1991) Simians, cyborgs, and women : the reinvention of nature. New York: Routledge.

Hart, C. (2008) 'Critical discourse analysis and metaphor: Toward a theoretical framework.', Critical Discourse Studies, 5(2), pp. 91–106.

Herzlich, C. (1973) Health and Illness: A Social Psychological Analysis. London: Academic Press.

Herzlich, C. and Pierret, J. (1987) Illness and Self in Society. Baltimore: Johns Hopkins University Press.

Heylighen, F. (2012) 'Heylighen, F., 2012. A Tale of Challenge, Adventure and Mystery: towards an agent-based unification of narrative and scientific models of behavior. Brussels, Belgium.' ECCO Working Paper Vrije Universiteit Brussel, Belgium. Available at: https://www.researchgate.net/publication/268265889_A_Tale_of_Challenge_Adventure_and_Mystery_towards_an_agent-based_unification_of_narrative_and_scientific_models_of_behavior (Accessed: 23 January 2024).

Hilts, D. (1982) 'TV Report On Vaccine Stirs Bitter Controversy', The Washington Post, 28 April.

Hobbs, A. (2013) 'The Deleterious Dominance of The Times in Nineteenth-Century Scholarship', Journal of Victorian Culture, 14(4), pp. 472–497.

Hobson-West, P. (2007) '"Trusting blindly can be the biggest risk of all": organised resistance to childhood vaccination in the UK', Sociology of Health & Illness, 29(2), pp. 198–215. Available at: https://doi.org/10.1111/j.1467-9566.2007.00544.x.

Hodgson (2022) 'Pathologising "Refusal": Prison, Health and Conscientious Objectors during the First World War.', Social history of medicine : the journal of the Society for the Social History of Medicine., 35(3), pp. 972–995.

Howarth, C. (2004) 'Re-presentation and resistance in the context of school exclusion: Reasons to be critical', Journal of Community and Applied Social Psychology, 14, pp. 356–377.

Howarth, C., Foster, J. and Dorrer, N. (2004) 'Exploring the potential of the theory of social representations in community-based health research – and vice versa?', Journal of Health Psychology, 9(2), pp. 229–243.

Iacobucci, G. (2024) 'Measles: Warning given over low MMR uptake after cases rise to 200 in West Midlands', BMJ, 384.

Jacoby, S. (2008) The age of American unreason : dumbing down and the future of democracy. London: Old Street.

Jahoda, G. (1988) 'Critical notes and reflections on "social representations"', European Journal of Social Psychology, 18(3), pp. 195–209. Available at: https://doi.org/10.1002/ejsp.2420180302.

Jodelet, D. (1991) Madness and Social Representations: Living with the Mad in One French Community. New Jersey: Prentice Hall.

Jovchelovitch, S. (1997) 'Peripheral communities and the transformation of social representations: Queries on power and recognition.', Social Psychological Review, 1(1), pp. 16–26.

Judt, T. (2005) Postwar: A History of Europe since 1945. New York: Penguin.

Kang, G.J. et al. (2017) 'Semantic network analysis of vaccine sentiment in online social media', Vaccine, 35(29), pp. 3621–3638. Available at: https://doi.org/10.1016/j.vaccine.2017.05.052.

Kim, L. and Kim, N. (2015) 'Connecting opinion, belief and value: semantic network analysis of a UK public survey on embryonic stem cell research', Journal of Science Communication, 14(1).

Kitta, A. (2012) Vaccinations and Public Concerns in History: Legend, Rumor and Risk Perception. New York: Routledge.

Kleinman, A. (2020) Illness Narratives : Suffering, Healing, and the Human Condition. New York: Basic Books.

Kress, G. and Van Leeuwen, T. (1996) Reading Images: The Grammar of Visual Design. 2nd edn. London: Routledge.

Kulenkampff M, Schwartzman JS, Wilson J. Neurological complications of pertussis inoculation. Arch Dis Child. 1974;49(1):46–49.

Lakoff, George. and Johnson, M. (1980) Metaphors we live by. Chicago ; University of Chicago Press.

Lamb, F. (1957) 'The People who Fight Polio', The Daily Telegraph, 26 September, p. 9.

Le Fanu, J. (1995) 'Victory over the plague of youth', The Times, 6 April, p. 13.

Leach, J. and Fairhead, M. (2007) Vaccine Anxieties: Global Science, Child Health and Society. London and New York: Routledge.

Leary, P. (2005) 'Googling the Victorians', Journal of Victorian Culture, 10(1), pp. 72–86. Available at: https://doi.org/10.3366/jvc.2005.10.1.72.

Lee-Potter, L. (1974) 'The questions we Must ask', Daily Mail, 25 September, p. 7.

Levy, M. (2020) 'If vaccines are a shot at freedom, why would anyone refuse?; The Covid-19 trial halted last week due to an "unexplained illness" has bolstered the antiinnoculation brigade', The Daily Telegraph (London), 14 September.

Leydesdorff, L. and Milojević, S. (2015) 'Scientometrics', in M. Lynch (ed.) International Encyclopedia of Social and Behavioral Sciences (2nd Edition). 2nd edn. Oxford: Elsevier, pp. 322–327.

Lintern, S. (2020) 'Coronavirus: UK lockdown could be indefinite until a vaccine is found, warn scientists advising government; "We are essentially waiting for a vaccine," experts say. "A vaccine is not five months away. We know it's at least 12 to 18 months away. So we will have difficult choices to make".', The Independent (United Kingdom), 16 March.

Livingstone, S. (2019) 'Audiences in an Age of Datafication: Critical Questions for Media Research', Television & New Media, 20(2), pp. 170–183. Available at: https://doi.org/10.1177/1527476418811118.

Loshak, D., Health Services Correspondent (1977) 'Whooping Cough Fear by Doctors', The Daily Telegraph, 22 July, p. 5.

Loshak, D., Health Services Correspondent and Our Medical Consultant (1977) 'MP blames vaccine for turning babies into cabbages?', The Daily Telegraph, 6 January, p. 2.

Luckhurst, T. (2016) '"The Vapourings of Empty Young Men?"', Journalism Studies, 17(4), pp. 475–489. Available at: https://doi.org/10.108 0/1461670X.2015.1071196.

Luo, C. et al. (2021) 'Exploring public perceptions of the COVID-19 vaccine online from a cultural perspective: Semantic network analysis of two social media platforms in the United States and China', Telematics and Informatics, 65, p. 101712. Available at: https://doi.org/10.1016/j.tele.2021.101712.

Lyu, J.C., Han, E.L. and Luli, G.K. (2021). 'COVID-19 Vaccine-Related Discussion on Twitter: Topic Modeling and Sentiment Analysis.' Journal of medical Internet research, 23(6).

MacDonald, N. (2015) 'Vaccine hesitancy: Definition, scope and determinants.' Vaccine, 33, 34, pp. 4161–4.

Madsen-Brooks, L. (2013) '"I Nevertheless Am a Historian": Digital Historical Practice and Malpractice around Black Confederate Soldiers', in J. Dougherty and K. Nawrotzki (eds) Writing History in the Digital Age. University of Michigan Press, pp. 49–63.

Mahrt, M. and Scharkow, M. (2013) 'The Value of Big Data in Digital Media Research', Journal of Broadcasting & Electronic Media, 57(1), pp. 20–33.

Malone, Carolyn (2002) 'Campaigning Journalism: The Clarion, The Daily Citizen, and the Protection of Women Workers, 1898-1912', Labour History Review, 67(3), pp. 281–297. Available at: https://doi.org/10.3828/lhr.67.3.281.

Martinez-Garcia, M., Camacho, J. and Hernández-Lemus, E. (2022) 'Connections and Biases in Health Equity and Culture Research: A Semantic Network Analysis', Frontiers in Public Health, 10, p. 834172. Available at: https://doi.org/10.3389/fpubh.2022.834172.

Maxwell-Stewart, H. (2016) 'Big Data and Australian History', Australian Historical Studies, 47(3), pp. 359–364. Available at: https://doi.org/10.1080/1031461X.2016.1208728.

Mayer-Schönberger, V. and Cukier, K. (2013) Big data: A revolution that will transform how we live, work, and think. Boston, MA: Houghton Mifflin Harcourt.

McKeever, B.W. et al. (2016) 'Silent Majority: Childhood Vaccinations and Antecedents to Communicative Action', Mass Communication and Society, 19(4), pp. 476–498. Available at: https://doi.org/10.1080/15205436.2016.1148172.

McKinnon, J.A. (1978) 'The impact of the Media on whooping cough immunization', Health Education Journal, 37(3), pp. 198–202. Available at: https://doi.org/10.1177/001789697803700307.

Merton, R.K. (1938) 'Social Structure and Anomie', American sociological review, 3(5), p. pp. 672–682.

Merton, R.K. (1973) The Sociology of Science: Theoretical and empirical investigations. Chicago/London: University of Chicago Press.

Milligan, I. (2013) 'Illusionary Order: Online Databases, Optical Character Recognition, and Canadian History, 1997-2010', Canadian Historical Review, 94(4), pp. 540–569.

Millward, G. (2016) 'A disability act? The Vaccine Damage Payments Act 1979 and the British government's response to the pertussis vaccine scare Social History of Medicine 30:2 429–47.

Millward, G. (2019) Vaccinating Britain: Mass vaccination and the public since the Second World War. Manchester: Manchester University Press.

Mold, A. (2015) Making the patient-consumer Patient organisations and health consumerism in Britain. Manchester, UK: Manchester University Press.

Moscovici, S. (2000) Social Representations: Explorations in Social Psychology. Edited by G. Duveen. Cambridge: Polity.

Musolff, A. (2012) 'The study of metaphor as part of critical discourse analysis', Critical Discourse Studies, 9(3), pp. 301–310. Available at: https://doi.org/10.1080/17405904.2012.688300.

Nicholson, B. (2013) 'The Digital Turn', Media History, 19(1), pp. 59–73.

Noelle-Neumann, E. (1974) 'The spiral of silence a theory of public opinion', Journal of Communication, 24(2), pp. 43–51.

Nuki, P. (2020) 'Search for a vaccine is this century's Space Race - and the stakes could not be higher', The Daily Telegraph (London). 11 April.

Nuwarda RF, Ramzan I, Weekes L, Kayser V. (2022) Vaccine Hesitancy: Contemporary Issues and Historical Background. Vaccines (Basel). Sep 22;10(10):1595. doi: https://doi.org/10.3390/vaccines10101595.

O'Halloran, K. (2007) 'Critical Discourse Analysis and the Corpus-informed Interpretation of Metaphor at the Register Level', Applied Linguistics, 28(1), pp. 1–24. Available at: https://doi.org/10.1093/applin/aml046.

Ong, J.C. and Cabanes, J.V.A. (2019) 'When Disinformation Studies Meets Production Studies: Social Identities and Moral Justifications in the Political Trolling Industry', International journal of communication (Online), p. 5771.

ONS (2023) 'Prevalence of ongoing symptoms following coronavirus (COVID-19) infection in the UK'. Office for National Statistics (ONS).

Oshinsky, D.M. (2006) Polio: An American Story New York: Oxford University Press

Our Own Correspondent (1955) 'Caution on Polio Vaccine Urged', The Daily Telegraph, 28 November, p. 9.

Our Own Correspondent (1955a) 'Growing Polio Virus: Method without Monkeys', The Daily Telegraph, 2 July, p. 7.

Our Own Correspondent (1955b) 'Polio Vaccine "Short Cut" Attempted', The Daily Telegraph, 12 May, p. 12.

Our Parliamentary Staff (1974) 'Vaccine Risk to Children?', The Daily Telegraph, 1 February, p. 10.

Our Political Correspondent (1957) 'Britain to Buy Salk Vaccine', The Daily Telegraph, 12 September, p. [1]+.

Our Special Correspondent (1956) 'Polio Risk in Taking Aspirin, Says Doctor', The Daily Telegraph, 11 July, p. 7.

Park, A. and Ducharme, J. (2021) 'Time Heroes of the Year 2021: THE MIRACLE WORKERS', Time, 13 December.

Parker, I. (1987) '"Social representations": Social psychology's (mis)use of sociology', Journal for the Theory of Social Behaviour, 17(4), pp. 447–469. Available at: https://doi.org/10.1111/j.1468-5914.1987.tb00108.x.

Parsons, T. (1951) The social system. London: Routledge & Kegan Paul.

Patterson, J.T. (1987) The dread disease: cancer and modern American culture. Cambridge, MA, Harvard University Press

Porat, T. et al. (2021) '"Vaccine Passports" May Backfire: Findings from a Cross-Sectional Study in the UK and Israel on Willingness to Get Vaccinated against COVID-19. 9, 902.', Vaccines [Preprint].

Potter, J. and Litton, I. (1985) 'Some problems underlying the theory of social representations.' British Journal of Social Psychology, 24, pp. 81–90.

Potts, A. and Semino, E. (2019) 'Cancer as a Metaphor', Metaphor and Symbol, 34(2), pp. 81–95. Available at: https://doi.org/10.1080/1092648 8.2019.1611723.

deSolla Price, D. (1965) 'Networks of Scientific Papers', Science, 149(3683), pp. 510–515.

Reisfield, G.M. and Wilson, G.R. (2004) 'Use of Metaphor in the Discourse on Cancer', Journal of Clinical Oncology, 22(19), pp. 4024–4027. Available at: https://doi.org/10.1200/JCO.2004.03.136.

Retter, E. (2022) 'Jab lies so nearly killed me; CORONAVIRUS CRISIS: PATIENT'S STARK WARNING', Daily Mirror, 7 January.

Ridley, S. (2021) 'I've seen deaths... but won't have jab.' The Sun (England), 20 June.

Roberts, D. (2021) 'Sunflower disability symbol "hijacked" by some anti-maskers.' South Wales Echo, 5 August.

Rose, S., Tuppen, S. and Drosopoulou, L. (2015) 'Writing a Big Data history of music', Early Music, 43(4), pp. 649–660. Available at: https://doi. org/10.1093/em/cav071.

Rosselli, R., Martini, M. and Bragazzi, N.L. (2016) 'The old and the new: vaccine hesitancy in the era of the Web 2.0. Challenges and opportunities.', J Prev Med Hyg., 57(1), pp. E47–50.

Rothman, S. M. (1995) Living in the Shadow of Death: Tuberculosis and the Social Experience of Illness in American History Baltimore: Johns Hopkins University Press

Ruiz, J. and Barnett, G.A. (2015) 'Exploring the presentation of HPV information online: A semantic network analysis of websites.' Vaccine, 33(29), pp. 3354–9.

Santavicca, T., Vanier-Clément, A. and Rousseau, C. (2022) 'Preventing and Appeasing COVID-19 Vaccine Tension in Schools to Protect the Well-Being of Children and Adolescents in Québec', International Journal of Child and Adolescent Resilience, 9(1).

Sass, E., Gottfried, G. and Sorem, A. (1996) Polio's Legacy. Lanham, Maryland: University Press of America.

Saunders, J. (2022) 'The making of "NHS staff" as a worker identity, 1948-85", in J. Crane and J. Hand (eds) Posters, Protests and Prescriptions: Cultural Histories of the National Health Service. Manchester, UK: Manchester University Press.

Schama, S. (2023) Foreign Bodies: Pandemics, Vaccines and the Health of Nations. New York: Simon & Schuster (epub.).

Silver, J. and Wilson, D. (2007) Polio Voices: An Oral History from the American Polio Epidemics and Worldwide Eradication Efforts. Connecticut: Praeger (The Praeger Series on Contemporary Health and Living).

Sjøvaag, H. and Karlsson, M. (2017) 'Rethinking Research Methods for Digital Journalism Studies', in B. Franklin and S.A. Eldridge II (eds) The Routledge

Companion to Digital Journalism Studies. London and New York: Routledge, pp. 87–95.

Small, N., Downs, M. and Froggatt, K. (2006) 'Improving end-of-life care for people with dementia – the benefits of combining UK approaches to palliative care and dementia care', in B. Miesen and G. Jones (eds) Care Giving in Dementia Research and applications. London: Routledge, pp. 365–392.

Smith, R.A. and Parrott, R.L. (2012) 'Mental representations of HPV in Appalachia: Gender, semantic network analysis, and knowledge gaps', Journal of Health Psychology, 17(6), pp. 917–928. Available at: https://doi.org/10.1177/1359105311428534.

Snowden, F.M. (2019) Epidemics and Society. Yale University Press. Available at: https://doi.org/10.2307/j.ctvqc6gg5.

Sontag, S. (1977) Illness as Metaphor. New York: Farrar, Straus and Giroux.

Sontag, S. (1989) Aids and its metaphors. New York: Farrar, Straus and Giroux.

Spitz, A. and Horvát, E.-Á. (2014) 'Measuring Long-Term Impact Based on Network Centrality: Unraveling Cinematic Citations', PLOS ONE, 9(10), p. e108857. Available at: https://doi.org/10.1371/journal.pone.0108857.

Stacey (1997) Teratologies: A Cultural Study of Cancer. London: Routledge.

Stahl, J.P., Cohen, R. and Denis, F. et al (2016) 'The impact of the web and social networks on vaccination. New challenges and opportunities offered to fight against vaccine hesitancy.' Med Mal Infect., 46(3), pp. 117–122.

Stead, W. (1888) 'Government by journalism', Contemporary Review, (49), pp. 653–657.

Steyerl, H. (2016) 'A Sea of Data: Apophenia and Pattern (Mis-)Recognition', E-Flux journal, p. online.

Storr, W. (2020) The Science of Storytelling: Why Stories Make Us Human, and How to Tell Them Better. London: William Collins.

Tanaka, S. (2013) 'Pasts in a Digital Age', in J. Dougherty and K. Nawrotzki (eds) Writing History in the Digital Age. University of Michigan Press, pp. 35–46.

Tarrow, S. (1993) 'Cycles of Collective Action: Between Moments of Madness and the Repertoire of Contention', Social Science History, 17(2), pp. 281–307. Available at: https://doi.org/10.2307/1171283.

The Daily Telegraph (1955) 'Polio Vaccine Supplies in Autumn', 26 April, p. 15.

Thomas, R. (2018) Advocacy Journalism in T.P. Vos (ed) Journalism Boston: Walter de Gruyter 391–414.

The Times (1937) 'Man Kept Alive In "Iron Lung"', 3 June, p. 13.

The Times (1938) 'Woman Of 26 Placed In Iron Lung', 3 August, p. 12.

The Times (1977) 'Not Really Reassuring', 15 June, p. 17.

Toscano, A. (2012) 'Seeing it whole: staging totality in social theory and art', The Sociological Review, 60(S1), pp. 64–83.

Trčková, D. (2015) 'Representations of Ebola and its victims in liberal American newspapers', Topics in Linguistics, 16(1), pp. 29–41. Available at: https://doi.org/10.2478/topling-2015-0009.

Tuchman, G. (1972) 'Objectivity as Strategic Ritual: An Examination of Newsmen's Notions of Objectivity', American Journal of Sociology, 77(4), pp. 660–679.

UKRI (2021) 'The story behind the Oxford-AstraZeneca COVID-19 vaccine success', 30 November. Available at: https://www.ukri.org/news-and-events/tackling-the-impact-of-covid-19/vaccines-and-treatments/the-story-behind-the-oxford-astrazeneca-covid-19-vaccine-success/ (Accessed: 23 January 2024).

Van Dijk, T. (1993) Discourse and Elite Racism. London: Sage.

Voelklein, C. and Howarth, C. (2005) 'A review of controversies about social representations theory: a British debate', Culture and Psychology, 11(4), pp. 431–454.

Wagner, W. et al. (1999) 'Theory and Method of Social Representations', Asian Journal of Social Psychology, 2. Available at: https://doi.org/10.1111/1467-839X.00028.

Wahl-Jorgensen, K. (2019) Emotions, Media and Politics. Cambridge: Polity.

Wald, P (2008) Contagious Cultures, Carriers, and the Outbreak Narrative Durham, NC US: Duke University Press.

Walkowitz, J. (1992) City of Dreadful Delight: Narratives of Sexual Danger in Late-Victorian London. Chicago/London: University of Chicago Press.

Wallis, P. and Nerlich, B. (n.d.) 'Wallis, Patrick and Brigitte Nerlich. "Disease metaphors in new epidemics: the UK media framing of the 2003 SARS epidemic." Social Science & Medicine (1982) 60 (2005): 2629–2639.

Weiner, J. (ed.) (1988) Papers for the Millions: the New Journalism in Britain, 1850s to 1914. Westport, Conn.: Greenwood Press.

Weiner, J. (2012) The Americanization of the British Press, 1830s-1914: Speed in the Age of Transatlantic Journalism. Basingstoke: Palgrave Macmillan.

Williams Camus, J.T. (2009) "Metaphors of cancer in scientific popularization articles in the British press." Discourse Studies 11 (2009): 465–495.

Wright Kennedy, S., Kuzmin, J.C. and Jones, B. (2017) 'New Methods in the History of Medicine: Streamlining Workflows to Enable Big-Data History Projects', Medical History, 61(3), pp. 477–480. Available at: https://doi.org/10.1017/mdh.2017.54.

Yoo, M., Lee, S. and Ha, T. (2019) 'Semantic network analysis for understanding user experiences of bipolar and depressive disorders on Reddit', Information Processing & Management, 56(4), pp. 1565–1575. Available at: https://doi.org/10.1016/j.ipm.2018.10.001.

Yoo, S. and Lim, G. (2021) 'Analysis of News Agenda Using Text mining and Semantic Network Analysis: Focused on COVID-19 Emotions,"', Journal of Intelligence and Information Systems., 27(1), pp. 47–64.

Index

© The Author(s), under exclusive licence to Springer Nature
Switzerland AG 2024
A. Cavanagh, *Anti-Vaccination and the Media*,
https://doi.org/10.1007/978-3-031-70559-5